20+ YEARS OF URBAN REBUILDING

Following the destruction of the World Trade Center and the surrounding area of Lower Manhattan from the terrorist attack of 9/11/2001 there were many heroic and extensive efforts to rebuild this iconic urban area in New York City. Political accomplishments, economic recovery, and community rehabilitation were urgent and important concerns and were continually monitored and debated. Supporting this progress and restoration of the critical infrastructure and built environment, however, was a vast and varied gathering of legislative bodies, public institutions, interest groups, and private individuals and entities. This is their story.

This book commences with a damage assessment of the immediate aftermath of the attack, describing the extent of destruction to the physical environment—buildings, public places, subway and train stations, roads and sidewalks—and the adverse consequences for the metropolitan economy, local businesses, communities, and families. Then, the story of the long route to recovery is presented, from early visionary intentions through brilliant leadership that confronted daunting bureaucratic procedures, to community voices achieving significant outcomes and eventually to an effective and exemplary partnership of public and private interests, that has produced the current vibrant urban center of downtown New York 23 years later.

Of particular interest to researchers, students, and practitioners of urban development and planning, *20+ Years of Urban Recovery* contributes to current research on the urban development crisis by focusing on the unique features in rebuilding urban centers following an unforeseen event of major devastation.

Patrice Derrington has over 20 years' experience on "Wall St.," leading high-level property finance and investment transactions. Combining architectural and development skills, she has also been involved in rebuilding efforts in inner-city neighborhoods and was the executive responsible for formulating a revitalization strategy for the devastated Lower Manhattan urban fabric after 9/11. Now serving as the Holliday Associate Professor and Director of the Real Estate Development program at Columbia University, she researches the urban development process and its analytical principles. She is the author of *Built Up: An Historical Perspective on the Contemporary Principles and Practices of Real Estate Development,* also published by Routledge (2021).

Rosemary Scanlon began her career in New York as an economist for the Port Authority of New York and New Jersey, serving as its Chief Economist from 1983 to 1993. She was New York State Deputy Comptroller from 1993 to 1997, with the assignment to monitor the budget and economy of New York City. From 1997 to 1999, Rosemary was a Visiting Research Fellow at the London School of Economics, where she recommended establishing a dedicated economics unit for the new mayoral government of London. She joined the New York University (NYU) Schack Institute of Real Estate in 2000 as Associate Professor of Real Estate Economics and became Divisional Dean from 2011 until retiring in 2015.

20+ YEARS OF URBAN REBUILDING

Lessons from the Revival of Lower Manhattan after 9/11

Patrice Derrington and Rosemary Scanlon

Routledge
Taylor & Francis Group
NEW YORK AND LONDON

Designed cover image: View of World Trade Center as rebuilt in 2023. Photograph by Joe Woolhead, courtesy of Silverstein Properties

First published 2024
by Routledge
605 Third Avenue, New York, NY 10158

and by Routledge
4 Park Square, Milton Park, Abingdon, Oxon, OX14 4RN

Routledge is an imprint of the Taylor & Francis Group, an informa business

ISBN: 9781032747040 (hbk)
ISBN: 9781032759098 (pbk)
ISBN: 9781003476139 (ebk)

DOI: 10.1201/9781003476139

Typeset in Zurich
by Newgen Publishing UK

For visionary real estate professionals of our future built world

Contents

Contents

Contents

Preface

Our goal for this discussion is to cast the net wider and look at the lower Manhattan neighborhood with a broader perspective of both geography and time ... to evaluate what sort of neighborhood it has become ... what kind of public realm it has, with a constant eye to the Memorial that exists at its center and to what role that plays in the new public realm that has developed inevitably. There are three layers of time: the neighborhood that existed when the World Trade Center began construction in 1965, the neighborhood that existed when the Twin Towers were destroyed in 2001, and the neighborhood that exists now. I think they're all different. (Paul Goldberger, Moderator of "The Memorials, the Public Realm, and Downtown as a 24-Hour City" discussion panel, May 9, 2017)

Purpose of this Book

Here, we seek to highlight the substantial progress in rebuilding the World Trade Center (WTC), together with the revitalization of the Lower Manhattan economy and urban fabric since the catastrophic impact of the terrorist attack of September 11, 2001.

Remarkable progress has been made since that day of tragedy and destruction. Not only have most of the buildings been rebuilt, but Lower Manhattan has been transformed over the past 20 years. By the 18th year of recovery, the neighborhood was bustling with vibrant activities: More than 14 million people visited the WTC Plaza, Memorial Pools, and the 9/11 Memorial & Museum, and there were almost 64,000 residents.

By 2019, a new workspace of almost 9 million square feet was built to accommodate the expanding private-sector employment of 250,815 people, with the addition of new technology and creative industries. The soaring new office buildings of 1, 3, and 4 WTC were completed and occupied, joining the successful completion of the first building, 7 WTC, in 2007.

The area had also become a culinary three-star destination as rated in Michelin's *Green Guide New York City* for 2019. Hotel rooms numbered almost 8,000, and new cultural and entertainment destinations opened, including the St. Nicolas Greek Orthodox Church in December 2022 and the Perelman Performing Arts Center in September 2023.

This astounding and impressive renewal of an important inner urban area was not achieved easily or readily, but rather was the result of a long process of complex political, economic, and social undertakings, often amidst conflict and burdened with the glaring risk of failure. These challenges arose at the outset, sometimes persisted tenaciously, and even transmogrified into new forms with new features, but, as we will show, they were mostly diligently and carefully resolved. In this story, the diligence, the overcoming of boundaries, and the determination to reach such an outstanding result provides numerous lessons in the arduous and complicated task of urban revival.

In addition to the extraordinary memorial and museum built on the footprints of the Twin Towers, the new Lower Manhattan boasts magnificent and efficient new train and subway hubs, spectacular new office towers, vast world-class retail and pedestrian areas, new apartment buildings, hotels, and restaurants, and a multitude of sites for visitors and tourists from near and far, all with state-of-the-art interconnectivity at multiple levels. Far from the quiet pre-9/11 office district that was deserted by 6 p.m., the area has become a vibrant 24/7 live-work-play zone with a larger population and a much improved quality of life. Further, the new infrastructure and transportation systems are designed to accommodate many future occupants and additional developments of apartments, offices, hotels, and restaurants.

At the adjacent former World Financial Center, now called Brookfield Place, the office buildings, the Winter Garden, and the retail complex have been either rebuilt or refurbished to a high contemporary standard, with new tenants for both the office space and the retail and restaurant locations.

Elsewhere in this area of downtown Manhattan, a large new terminal has opened for the Staten Island Ferry; new water, sewer, electrical lines, and telecommunication cables have been laid below the streets; the Fulton Transit Center is open and thriving with new retail shops serving the thousands of daily passengers who now connect between the six major subway lines and through glistening, white-tiled tunnels to the Port Authority Trans-Hudson (PATH) line at the WTC. The Cortlandt Street station on the R subway line has reopened, the WTC Cortlandt station on the #1 Subway line has been completely reconstructed, and most recently, the South Ferry station has been rebuilt for a second time—this following the further damage of seawater infiltration during the hurricane Superstorm Sandy on October 29, 2012. The infrastructure work throughout Lower Manhattan to overcome the destruction of both the 9/11 and the superstorm events has been Herculean, accomplished by professional engineers and planners, consultants, and skilled workers from the many agencies involved in the reconstruction and new building.

Then, progress was abruptly halted in March 2020 by the Coronavirus pandemic as people left their offices, sequestered themselves in their residences, and the characteristics of everyday life in the emerging Lower Manhattan simply shut down. Not until late 2021 and early 2022 were there again signs of rebuilding and recovery from the successive catastrophes.

How did this successful rebuilding of a substantial area in a major global city happen, and what can we learn from the dramatic transformation of the neighborhood? Is the new Lower Manhattan only a result of the unique, tragic circumstances of 9/11, or does it provide possible blueprints and direction for reimagining and redeveloping other urban centers and for city-building projects everywhere? Also, we ask: Given what we have learned in rebuilding Lower Manhattan and given the technological and sociological developments since 2001, if we had to do it all over again, what would we change, and what would we do differently?

To answer these questions, at Columbia University's Center for Urban Real Estate, we studied reports from diverse sources and sought out the many capable players involved over the past 20 years. They represent a rich tapestry of wide-ranging interests, diverse expertise, and a multitude of disciplines: commercial developers, government officials, community groups, nonprofit entities, architects, engineers, urban planners, lawyers, politicians, journalists, transportation experts, residents, media critics, and museum specialists. Each has brought their own motivations and incentive structures, but all shared the overarching goal to rebuild something we could all be proud of.

To mark the first 15 years of rebuilding, we brought many of these voices together in a series of panel discussions at a five-series symposium we called "T+15." These presentations and debates involved government officials, professionals, and scholars so that the benefit of their experience, the human reactions, the community dynamics, and the needs of urban systems could be shared.[1]

Combining the insights of the T+15 symposium with reports and published reviews, we present the major, and sometimes hidden, challenges of rebuilding the WTC. Included and interwoven are the concerns for the Lower Manhattan economy, the hurdles and highlights of the planning process, and the various visions for this important urban area put forth by the major stakeholders and citizenry. We also examine the unprecedented technical and logistical task of rebuilding the infrastructure and key transportation arteries and hubs, the revisioning of the public realm of the WTC site, and the creative genius of the Memorial, the Museum, the Oculus, and other impressive buildings, along with the questioning of the commercial rationale in replacing 10 million square feet of office space and substantial volumes of retail space and the likelihood of attracting new residents in the aftermath of the catastrophe.

We provide a timeline of this inspiring yet complex redevelopment and present the visions, decisions, and actions of major stakeholders—federal, state, and local governments, and the Port Authority of New York and New Jersey (PANYNJ) as the owner of the 16-acre WTC site. We include the determination and decisions of Silverstein Properties, which, just six weeks before the terrorist attack, had agreed to a 99-year lease of the office properties on the site. This narration is enriched with interviews, called "Reflections," from many of the major principals, planners, architects, engineers, and community representatives who were responsible for the rebuilding. For this book, they have shared their personal recollections of key events as well as their assessments of the challenges and accomplishments of rebuilding the WTC and transforming the economy of Lower Manhattan.

We hope that lessons emerging from this review and the insightful reflections will guide future endeavors in urban rebuilding in other global cities. While we may touch on academic theory and history, the symposium and subsequent compilation of the ideas and reflections were designed to be practical and forward-looking. Recognizing that the world is undergoing rapid urbanization, challenging existing municipal, spatial, and functional fundamentals, we planned these discussions around constructive procedural and managerial responses for understanding and shaping future urban environments.

Together with some exemplary photographs taken by Joe Woolhead for Silverstein Properties, beginning on the day of the terrorist attack and lasting throughout the rebuilding process, the authors offer this book—this testament of successful rebuilding—to the people of New York City and the metropolitan area, and to keen observers of cities throughout the world.

Note

1 See Appendix 1 for a list of these panel discussions and the names of the panelists.

Acknowledgements

First and foremost, we extend our thanks to the 32 developers, architects, government officials, industry leaders, and all observers of the remarkable rebuilding of the World Trade Center after the catastrophic terrorist attack on the Twin Towers on September 11, 2001. Their insightful and heartfelt stories are included here in their quotations, their contributions to panel discussions, and especially in our series of Reflections, where each story is presented in their own words.

Many of these writers of Reflections participated in our symposium comprising five panel discussions, held between late 2016 and mid-2017 (see symposium details in Appendix I). The issues raised, the details discussed, and the assessments of the rebuilding gave us the initial thrust of this book. We are immensely grateful for their participation and insights.

Also contributing an initial curiosity that spurred the early discussions and detailed research on the intersection of the public and private sectors was Robin Panovka, the Co-Chair of Real Estate and REIT (Real Estate Investment Trusts) M&A (mergers and acquisitions) practices at Wachtell, Lipton, Rosen & Katz, an eminent New York legal firm that has been involved in many aspects of the project over the past 20+ years. The necessity, challenges, and achievements of this unusual cooperation between public authorities and private real estate interests is a perspective on the rebuilding process that we have taken as being of key importance.

For supporting the production of the symposium and research for this book, we are grateful to Columbia University's Center for Urban Real Estate and the generous gift by Johnny Din.

In achieving the compelling visual references for our story, we are particularly grateful to the photographer Joe Woolhead, who documented the devastation of 9/11 and so much of the rebuilding progress throughout the decades since. We also express our gratitude to Dara McQuillen, Chief Marketing & Communications Officer at Silverstein Properties, Inc. for guiding us through the remarkable range of Joe's photographic history. Additionally, we are thankful to numerous other people and entities for contributing important images.

We would also like to thank Ed Needle, the Engineering Publisher at Routledge/CRC Press/Taylor & Francis Group, for his interest in our topic and support for the book; and for the patient, detailed, and helpful advice at Routledge from Megan Hiatt, Senior

Acknowledgements

Production Editor, Chelsea Reeves, Senior Editorial Assistant, Kirsty Hardwick, Editorial Assistant; Newgen Publishing's Kelly Winter, Project Manager, and Helen Lund, a most meticulous copyeditor, in preparing this book for publication.

Author's Note: Throughout this book, the geographic focus of our story is variously referred to as "Lower Manhattan" or as "Downtown Manhattan." Both terms are interchangeable and refer to the area of Manhattan south of Canal Street through to The Battery at the southern tip of the island of Manhattan.

About the Authors

0.1

Patrice Derrington.

Holliday Associate Professor of Real Estate

Columbia University in the City of New York

Patrice Derrington has over 20 years' experience on "Wall St.," leading high-level property finance and investment transactions. Combining architectural and development skills, she has also been involved in rebuilding efforts in inner-city neighborhoods and was the executive responsible for formulating a revitalization strategy for the devastated Lower Manhattan urban fabric after 9/11. Now serving as the Holliday Associate Professor and Director of the Real Estate Development program at Columbia University, she researches the urban development process and its analytical principles. She is the author of *Built Up: An Historical Perspective on the Contemporary Principles and Practices of Real Estate Development,* also published by Routledge (2021).

About the Authors

Rosemary Scanlon began her career in New York as an economist for the Port Authority of New York and New Jersey, serving as its Chief Economist from 1983 to 1993. She was New York State Deputy Comptroller from 1993 to 1997, with the assignment to monitor the budget and economy of New York City. Between 1997 and 1999, Rosemary was a Visiting Research Fellow at the London School of Economics, where she recommended establishing a dedicated economics unit for the new mayoral government of London. She joined the New York University (NYU) Schack Institute of Real Estate in 2000 as Associate Professor of Real Estate Economics and became Divisional Dean from 2011 until retiring in 2015.

0.2

Rosemary Scanlon.

Divisional Dean (Retired)

Schack Institute of Real Estate, New York University

Chapter 1: Some Important History of Lower Manhattan

Manhattan is one of five boroughs of New York City (NYC). During the first half of the 20th century, its skyline of recognizable towers included the Empire State Building, the Chrysler Building, and Rockefeller Center, all located in Midtown Manhattan. However, in the 1970s, a new iconic addition was made to the southern tip in the form of the World Trade Center (WTC) with its soaring twin towers of office space.

New Commercial Space

The transformation of Lower Manhattan from an aging financial and business district with office buildings dating from the early 1900s to the 1920s began in the decade between 1947 and 1957. Alex Garvin, who would be appointed Head Planner at the New York State (NYS) agency responsible for the WTC rebuilding, explains the importance of this initial foray:[1]

> In the decade between 1947 and 1956, 15.1 million square feet of new office space—more than existed in the Loop in Chicago, the nation's second-largest office district—was erected in Midtown.... During that time, only 1.1 million square feet of office space was erected in Lower Manhattan in three modest office structures and one six-story addition.[2] The situation was so serious that in 1952, the *Journal of Commerce* published articles predicting that businesses would soon relocate from Lower Manhattan to Midtown.[3]
>
> The attempt to revive Lower Manhattan is central to an understanding of the World Trade Center. The first significant post-war revival effort came in November of 1955, when the Chase Manhattan Bank (under the leadership of David Rockefeller) announced that it would consolidate its nine-building, 8,700-employee operations on the two blocks bounded by Nassau, Liberty, William, and Pine streets. The scheme that emerged was a superblock designed by the firm of Skidmore, Owings & Merrill. A one-block section of Cedar Street was closed, creating a 2.5-acre site to be shared by a new 60-story, 1.7 million-square-foot office slab; the existing 38-story building that had housed Chase headquarters since 1928; and a new, spacious plaza.
>
> Between 1960, when the new Chase Manhattan Bank Building was ready for occupancy, and 1972, the year before the World Trade Center was officially

DOI: 10.1201/9781003476139-1

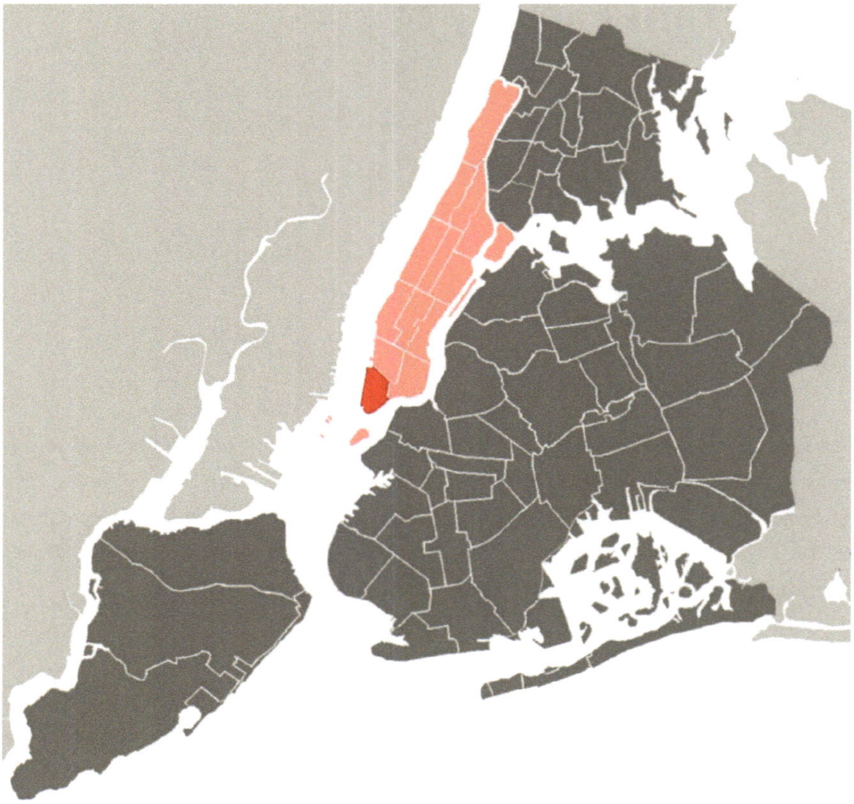

1.1
Lower Manhattan location map.
Source: New York State government.

dedicated, 45 buildings containing 32.5 million square feet of office space opened in Lower Manhattan.[4] However, these buildings did little to enliven the area.

In 1958, following Chase's commitment to Downtown, David Rockefeller led the formation of the Downtown Lower Manhattan Association (D-LMA). This was a membership organization of business executives that advocated for the revival of Lower Manhattan with new office buildings and public policy initiatives supporting the updating of the urban fabric.

Port Authority of New York and New Jersey

A key public contribution to this transformation was the agreement between New York (NY) State's Governor Nelson Rockefeller and Governor Hughes of New Jersey (NJ) to commit their bi-state agency, the Port Authority of New York and New Jersey (PANYNJ or Port Authority), to building the WTC in Lower Manhattan. For additional regional benefit, commercial development in this prime location would attract revenues to support the Port Authority's extensive infrastructure of ports, trains, bridges, and tunnels throughout NY and NJ. Since office space was at that time considered the highest and best use of land in this area to achieve maximum cash flow, the site was designated for that use. By the late 1960s, the construction of the WTC was underway.

With this new development, Lower Manhattan was competing as an important business center, as skyscrapers were rising along Park Avenue in New York's Midtown. Alex Garvin also describes the accomplishment:[5]

> The Port Authority changed the face of Lower Manhattan when it replaced 16 acres occupied by a few hundred businesses with the World Trade Center, which served tens of thousands of tenants and provided 12 million square feet of office space in 1, 2, 4, 5, and 7 World Trade Center. Seven hundred thousand square feet of office space was devoted to the U.S. Customs House, 600,000 square feet to the Vista Hotel, and 2.4 million square feet to the retail concourse, parking, services, and the PATH terminal. Notably, the PATH railway brought 50,000 to 60,000 commuters to Lower Manhattan on a daily basis.

This rapid growth needed to be directed. Robert Douglass, another visionary for the revival of Downtown Manhattan, became chair of the D-LMA and, through his lifelong advocacy, spurred the formation in 1993 of a business improvement district (BID), the Alliance for Downtown New York (the Downtown Alliance). According to a tribute posted on the D-LMA website (https://d-lma.com/) on December 7, 2016, Douglass "worked assiduously to make Lower Manhattan a premier address for businesses, residents, and visitors" and was hailed for his civic accomplishments.

The year 1973 proved to be pivotal for the transformation of Lower Manhattan when a truck fell through the elevated West Side Highway, bringing focus to this dilapidated but vital highway link between Downtown and Midtown. The replacement plan to build an underwater vehicular tunnel from this Downtown area up to 42nd Street (the "Westway Project") failed after a long series of court battles. However, those legal battles brought attention to the value of easy pedestrian access to the Hudson River waterfront, the importance of transport connections between Lower Manhattan and Midtown, and the need for thoughtful public policy to shape the use of the waterfront and pedestrian bridges.

By the mid-1970s, the 10 million square feet of office space at the WTC were completed. In addition, during that decade, another 22 million square feet of offices, including the new headquarters for the Federal Reserve Bank of New York, were located along the East River, on The Battery shorefront, along Water Street, and at the revived South Street Seaport. These modern space options were able to accommodate 15 percent of NYC's white-collar workforce.

Adding Residential and Public Space

During the 20th century, housing constituted a small proportion of the building fabric in Lower Manhattan. It had consisted solely of some small-scale, modest, and often historical residential buildings on the eastern side adjacent to the Brooklyn Bridge, along with two high-rise apartment towers completed in the early 1960s.

1.2

Original WTC twin towers.

Source: Jeff Mock, courtesy of Silverstein Properties.

Battery Park City

During the late 1960s, using fill from the excavation for the WTC, 92 acres of reclaimed land was established to the west along the Hudson River. This landfill was developed as housing with necessary residential and retail amenities, although it also had some office facilities for an NYS public-benefit corporation, the Hugh L. Carey Battery Park City Authority (BPCA). It became known as Battery Park City.

In the 1980s, with the vision of Paul Reichman and his Canadian development company, Olympia & York, and the imaginative architecture of César Pelli, the World Financial Center (WFC), now called Brookfield Place, attracted major financial institutions as tenants. Its exquisite glass-domed Winter Garden, with its array of exotic palm trees, became a major cultural magnet for Downtown and delighted tourists and locals through September 10, 2001.

During the 1980s, Battery Park was expanded with 13 new buildings to the south and a new home for Stuyvesant High School. This was followed by nine more buildings in the 1990s. In connection with the later rebuilding after 9/11, that extension of urban mixed-use living space was overseen by Amanda Burden, who would be the Director of the New York City (NYC) Planning Department in the Bloomberg administration and an important advocate in planning the rebuilding of the WTC site.

The completed project not only had 9.3 million square feet of new commercial buildings and 7.2 million square feet of urban housing, but also 36 acres of open space within the complex, a rare amenity in Manhattan.

1.3
World Financial Center, now Brookfield Place.
Source: Image courtesy of Brookfield Properties.

Although the development at Battery Park City had not been specifically intended to provide affordable housing, the earliest units were mostly affordable for average city workers, and in those early years, the residential complex represented a diverse socio-economic population. However, as Wall Street financial services expanded and required numerous new office towers, the moderately priced housing was taken up by high-earning financial service workers. The residential buildings were particularly suited to two-career families, although in 1998, to restore the socio-economic residential balance, the BPCA began providing subsidies to households with annual incomes below $108,000.

By 2000, the residential population of Downtown Manhattan was 20,088.[6]

Cultural Life

The burgeoning cultural life of the Downtown area, however, had begun even earlier as artists of the 1970s moved into abandoned warehouse buildings around the neglected South Street Seaport. Anita Contini[7] reflects:

> I think my love affair with Lower Manhattan began in the 1970s when I moved down here. It was an area where a lot of artists lived. They worked and lived side by side with the commercial core. They picked up most of the empty spaces and created studios ... Jack Youngerman, Ellsworth Kelly, Mark Desuvero, or Agnes Martin. It was an exciting place because they would gather in a café or meet on the side of the street. I was in a loft building on Front Street; it didn't have a front door, but I had 3,500 square feet on three floors, and the rent was $47. There were no grocery stores, so I would head off to Chinatown to find groceries. I would take my son to

1.4
Battery Park City.
Source: Courtesy of CITYNEIGHBORHOODS, NYC.

play with friends anywhere north of us since there weren't any other children down here at that time. But in the 1970s and 1980s, there was a lot going on related to the arts and culture. When I ran Creative Time, we had an exhibition called Ruckus Manhattan, which the artist Red Grooms built, and there were lines around the block. If it wasn't for all the arts and cultural activities that were happening in Lower Manhattan, I don't know if excitement (for downtown) would have built so quickly.

Vulnerability of the Vibrant Downtown

Despite its economic vitality in the post-war decades, the recession of the early 1990s brought severe setbacks to New York's real estate market and Lower Manhattan in particular. By 1994, vast quantities of space in the older pre-World War II buildings stood empty.

I remember in 1991, I was walking down William Street with a downtown real estate broker, and he was pointing out to me how many buildings were empty; in his judgment, they were never, ever going to be filled again. And across all of Downtown, including the newer buildings, one out of every four floors of office space was vacant. Also it was a ghost town down here after 5 or 6 p.m.; there were only a few hundred people living here. People were afraid that New York City, which was the only city in the country that had two central business districts, might end up with one. (Charles Bagli, journalist for *The New York Times*, Moderator, T+15 symposium session on "The Memorials, the Public Realm, and Downtown as a 24-Hour City," May 9, 2017)

Reviving Lower Manhattan through the Decade Prior to September 11, 2001

The first terrorist bombing of the WTC on February 26, 1993, killed six people and destroyed the basement area and lobby of 1 WTC as well as the lobby area of the adjacent Vista Hotel. An estimated 50,000 people were able to evacuate safely in a process that took several hours, although a number of people were trapped in stalled elevators during those many hours. Fortunately, as Eugene Fusillo, then Chief Engineer of the Port Authority,[8] conveyed to this book's co-author Rosemary Scanlon later in 1993, substantial destruction was avoided: "Were this bomb placed as accurately on any other large building in Manhattan, I believe it would have taken that building down. But, the Trade Center was highly engineered and well-rooted in bedrock."

Although it eventually recovered from the 1993 WTC bombing and the recession of 1990–1994, Lower Manhattan again felt the burden of its obsolete building stock. Now, David Rockefeller, as Emeritus Chair of the Chase Manhattan Bank and wishing to maintain the economy of the area, joined with Robert Douglass, Chair of the D-LMA and The Alliance for Downtown New York, a newly established business improvement district (BID), to seek further policy initiatives in support of revitalization. They and the inaugural President of the Downtown Alliance, Carl Weisbrod, worked closely with newly elected NYC Mayor Rudy Giuliani to pass a series of policy measures to reinvigorate the aging office buildings. A key action of these proposals was the passage in 1995 of a new NYC local law, 421-g, which gave financial incentives and a 30-year tax abatement for the conversion of older office buildings for residential development.

> In 1995, no one was renting the older office buildings. Everyone was looking for open space and (large) trading floors. Donald Trump bought 40 Wall Street for $8 million—$8 per square foot. And one of the biggest mistakes of my life was talking to AIG when they were selling it [their building] for $1 per square foot in 1990. They couldn't give it away. 421-g gave great tax credits and abatement if you renovated the old buildings. The old buildings were perfect for residential [use] because some of the slimmer ones had great light and air. 421-g was critical, changing Downtown to a 24/7 community. (Frank Sciame, CEO of Sciame Construction, panelist, T+15 symposium session on "The Memorials, the Public Realm, and Downtown as a 24-Hour City," May 9, 2017)

Changing Resident Demographics

Good design, as in César Pelli's World Financial Center, and thoughtful public policy to put the elite Stuyvesant High School downtown, and then to complement

> that with a public school in Battery Park City, helped to change the supply of basic services as well as the image of Downtown. The reduction in crime, initiated by Mayor Dinkins in the early 1990s, began to improve the life of the city's residents as well as change the image of New York City to the nation and the world. (Hugh Kelly, real estate economist, panelist, T+15 symposium session on "The Memorials, the Public Realm, and Downtown as a 24-Hour City," May 9, 2017)

As the urban conditions of NYC improved, Lower Manhattan attracted more residents and was becoming a lively 24/7 urban environment. The driving force behind this was a rare combination of municipal action and progressive corporate intentions.

Planning Strategy for Urban Residential Neighborhoods

In the panel discussion on May 18, 2017, former NYC Planning Department Director Carl Weisbrod related the changes that had occurred with respect to residential activity in Lower Manhattan in the decades prior to 9/11, including his tenure in the Planning Department. We were reminded that the broad decision for Lower Manhattan to be mixed-use emerged in those earlier decades.

In the early 1990s, Lower Manhattan's office vacancy had risen to 25 percent plus as many corporations and professional firms relocated to Midtown Manhattan. This was a consequence of the aging office building stock, few remaining sites for development, and the increasing supply of new office buildings in the western area of Midtown, particularly around the revitalized and rezoned Times Square neighborhood (Morgan Stanley, Nasdaq, etc.). This, in itself, was a lesson in the urban dynamic of cyclical trends and the unintended consequences for other urban areas when revitalization incentives are provided. This was exacerbated by the interrelated changes in commuter patterns in Manhattan. Commuters from the NYC suburbs, Northern New Jersey, and nearby Connecticut were choosing to complete their railway journeys in Midtown at the Penn and Grand Central stations, discarding the additional leg on the crowded and dilapidated subway lines to Lower Manhattan. The location of offices in Midtown was thus overwhelmingly preferred across the spectrum of white-collar workers.

Therefore, for office property, Lower Manhattan suffered a number of disadvantages:

- Aging buildings resulting in inferior building quality.
- Obsolete office environments with lesser floor-to-floor heights to accommodate raised floors for technology.
- Less convenience for commuters (this was before Brooklyn began its recent development of large multifamily properties that accommodated workers with a swift commute to Lower Manhattan).
- Filled mostly with office buildings, which had little population in the evenings and weekends, Lower Manhattan was regarded as a dull and boring place to be and attracted few tourists and visitors.[9]

Private Action to Expand Residential Living

Bill Rudin and Robert Douglass joined with the Downtown Alliance to consider what sort of business district Lower Manhattan should be and whether residential activity could be included. The issue was bitterly fought:

- Existing residents enjoyed their uncrowded conditions in the evenings and on weekends.
- Increased development would be disruptive to existing residents: increased garbage haulers and construction workers.
- To resolve this, existing residents sought the support of the municipality to resolve the conflict and discourage increases in residential living.
- But more NYC-wide housing was critically needed, and land values in Lower Manhattan were now lower than in Midtown and on the Upper East Side and Upper West Side.

Reflections of Carl Weisbrod[10]

The Alliance for Downtown New York started in 1994, partly as a consequence of the depth of despair that existed after the first bombing of the World Trade Center in February of 1993 but also because of the depression-like state of the office market in Lower Manhattan in the early 1990s, where the vacancy rate was 25 percent.

With the support of Mayor Giuliani, we set in motion a plan of action and a new tax policy called 421-g. There was no tax on buildings, perhaps for 15 years, only on land, and that was capped at a low level. 421-g triggered a residential boom in two ways. First, it greatly incentivized the transformation of older office buildings to residential use. These were the deepest real estate tax incentives ever in NYC.

Second, the legislation released all the obsolete office buildings for transactions. At the time 421-g went into place, those old buildings had been frozen in time for decades or more. The insurance companies had taken them over through foreclosures during the 1990–1995 recession, and they were kept out of use or at a low level of use so that their estimated valuation would not be adversely affected by the reality of new transactions at market rental rates.

So, at the Alliance, we had a strategy to transform Lower Manhattan from a one-dimensional business community to a multidimensional residential and business community. These older unusable office buildings became residential buildings and highly desirable residences. That was a joint effort of the downtown business community and the downtown residential community. I can't emphasize enough the importance of the alliance that existed and continues to exist between the downtown commercial sector and the downtown residential community, which has been a critical part of the redevelopment of Lower Manhattan.

> As a result of these policies and actions, on the day before 9/11, Downtown's re-emergence was well underway, if not almost completed. Lower Manhattan was probably in the best shape that it had been in decades. The World Trade Center was full. The commercial vacancy rate was actually lower in Downtown Manhattan than it was in Midtown Manhattan. Though, what 421-g didn't do, nor the later Liberty Bonds used for the rebuilding, was to create substantial affordable housing in Downtown. The Liberty Bonds delivered only 5 percent of the units as affordable.

The results of 421-g were a turning point for residential living in Lower Manhattan:

- The first development was at 20 Broad Street, and the second was at 45 Wall Street.
- The Downtown Alliance worked with the existing Community Board 1 to include the residential sector in their mandate, and 10 percent of the Downtown Alliance board was to be appointed by that community entity.
- Lower Manhattan was one of the first transformations in the United States of a commercial central business district (CBD) to include such substantial residential buildings.

> I happen to have the numbers on 421-g. We're talking about 7,500 apartments (renovated) between 1995 and 2006. That's a huge change, and 421-g absorbed an enormous portion of the old buildings. In 1970, there were 823 people in 257 apartments. Today, there are close to 60,000 people. (Alex Garvin, panelist, T+ 15 symposium session on "The Memorials, the Public Realm, and Downtown as a 24-Hour City," May 9, 2017)

Retail

With the growing population of office workers in the area, the need for retail space increased in Lower Manhattan. Particularly at the WTC, where not only office workers but also the PATH commuters traversed the center's lower arcades, the Port Authority gave strategic attention to retail activity in this area. The approximately 400,000 square feet of available retail space included a medley of services, including food services. But in those early days, this retail space did not appear to retain shoppers in the vicinity nor bring visitors to the commercial area outside of office hours, though significant exceptions included the restaurant Windows on the World in 1 WTC and the WTC Observatory on the top floors of 2 WTC.

In the 1990s, however, the repair of the WTC after the 1993 bombing and the repositioning of the retail and public space in nearby Brookfield Place led to a markedly improved quality of retail stores by the end of that decade:

> What many people don't realize is that after the 1993 bombing of the World Trade Center, we adjusted the retail environment at the WTC to respond to

> the [office] tenants that we had, as well as the tenants that were at the World Financial Center. We redeveloped the open space that was the large Austin Tobin Plaza. So, over a span, I think the "city within the city" of the World Trade Center has, in fact, responded to the neighborhood and adapted to the neighborhood and helped it thrive. And that's because of the [improvement] to transportation, commercial development, and public space. (Carla Bonacci, Chief Architect at the Port Authority at the time, panelist, T+15 symposium session on "The Memorials, the Public Realm, and Downtown as a 24-Hour City," May 9, 2017)

The Transforming Economy of New York City and Lower Manhattan

The larger New York City economy was also in transition during the 1980s and early 1990s. In 1991, Mitchell Moss, a Professor of Urban Policy at New York University, and Hugh O'Neill, Founder and President of his consulting group Appleseed, wrote a seminal research paper that documented the change underway in the Manhattan and metropolitan economy: "After decades of depending on manufacturing and goods-handling industries, a new economic platform was emerging, led by professional business services, the arts and culture, and initial corners of the new computer-based functions and job categories."[11]

By 2000, private-sector employment in Lower Manhattan had reached an estimated 265,000, following a strong economic recovery throughout NYC after the recession of the early to mid-1990s. While the financial services sector recovered solidly, as foreseen by Moss and O'Neill, Lower Manhattan continued to experience migration of the increasingly technology-assisted businesses to the more contemporary office space in Midtown Manhattan. But also, as the research noted, Manhattan's economic base shifted further from corporate headquarters to professional and commercial services, and many of these took office space in refurbished older buildings, particularly those Downtown, such as the "reclaimed" 10 Hanover Square.

> A developer had bought the old office building at 10 Hanover Square, and the carpenters were there literally framing out the building for apartments when Goldman Sachs walks up, knocks on the door, and says, "I think we'll take a floor here." And by the end of the conversation, they took the whole building; it immediately went back to commercial use as an office building. (Charles Bagli, journalist for *The New York Times*, Moderator, T+15 symposium session on "The Memorials, the Public Realm, and Downtown as a 24-Hour City," May 9, 2017)

Notes

1 Alex Garvin, "Negotiating the Mega-Rebuilding Deal at the World Trade Center: The Historical Context," *Transactions: The Tennessee Journal of Business Law* 10 (2008), pp. 15–26; quoted extract, pp. 15–16.

2 The Real Estate Board of New York, Inc., Research Department, *Office Building Construction: Manhattan 1947–1967* (1964).

3 Robert A. M. Stern, Thomas Mellins & David Fishman, *New York 1960: Architecture and Urbanism between the Second World War and the Bicentennial* (The Monticello Press, 1995), pp. 167–170.

4 NYC Department of City Planning, *Manhattan Office Space Market 1960–1978* (1979) (unpublished, on file with author).

5 Garvin, "Negotiating the Mega-Rebuilding Deal," p. 16.

6 U.S. Census Bureau (2000).

7 From 2002–2005, Anita Contini was Vice President and Director of Memorial and Cultural Programs at the LMDC, where she led the important worldwide competition for design of the WTC Memorial.

8 Gene, as he was known, was trapped in an elevator in 1 WTC, and using his house and car keys and his knowledge of how the Trade Center was built, gouged a hole in the elevator door large enough to crawl out and walk down the many flights of stairs.

9 Alliance for Downtown New York, *Lower Manhattan Real Estate Year in Review 1994*.

10 Carl Weisbrod was the first President of the Alliance for Downtown New York.

11 Mitchell Moss and Hugh O'Neill, *Reinventing New York: Competing in the Next Century's Global Economy* (Urban Research Center, Robin F. Wagner Graduate School of Public Service, New York University, 1991), p. 4.

Chapter 2: September 11, 2001

The terrorist attack on the WTC on the morning of September 11, 2001, caused the death of almost 3,000 people and the complete destruction of the Twin Towers, comprising 13 million square feet of office and retail space. In the surrounding neighborhood, an additional 15 million square feet of space was either effectively destroyed or severely damaged, the latter requiring months of reconstruction to fully restore it.

Further Reflections of Anita Contini

On September 11, I was in Washington, DC, giving a talk at the Hirshhorn Museum when I heard of the terrorist attack, its scale, and horror. My first thoughts were of my son, who often passed through the World Trade Center on his way to work, and of my husband, whose office was immediately across the street from the Trade Center's North Tower. I will never forget the hours of panic until I heard their voices and knew they were all right. Throughout my time involved with the Memorial, not a day went by when I did not think of the panic and pain of so many others, both victims and families, who were not so lucky. During the Memorial process, what sustained me and every member of the jury and all those involved was the hope and belief that we were part of a process that would help bring relief to those who lost so much and celebrate those whom they loved.

The first World Trade Center terrorist attack in 1993 was a tragic and jarring event whose largely local impact was, nevertheless, an early catalyst of discussion about the future of downtown as a whole.

The vastly more devastating and tragic attack on September 11, 2001, affected countless more individuals and families, all of Lower Manhattan, the city, the country, and the world. It was an event that demanded a response commensurate with its scale and impact. How to properly memorialize the attack, including those who came to help in the recovery efforts, was a critical challenge among a sea of others that Lower Manhattan faced. Of course, there were as many opinions as there were needs.

DOI: 10.1201/9781003476139-2

2.1
WTC collapses, September 11, 2001.
Source: Joe Woolhead.

Reflections of Alice M. Greenwald

On the morning of the 9/11 attacks, I was unpacking moving boxes in my new home in Washington, DC, having relocated only days before to assume the role of Associate Museum Director for Museum Programs at the United States Holocaust Memorial Museum. Like everyone on that terrible day, I was in a state of shock, watching in disbelief as I saw on my television screen, in realtime, the crash of hijacked Flight 175 into the South Tower at the World Trade Center. There was a carpenter working in the house, a contractor I barely knew; our mutual, spontaneous impulse, as we each struggled to comprehend what we were witnessing, was to reach for one another. Connecting with something human, with compassion in a moment of unimaginable grief, was the only thing in our power at that moment.

The news of the Pentagon attack and, later, reports of the crash of another hijacked plane in central Pennsylvania confirmed what we already had surmised: Our nation was under attack. And then, in front of our eyes, we watched as the South Tower collapsed. I remember crying out loud, "Oh, the people, the people," as I thought of all [those] inside those doomed buildings. I desperately tried to reach the office at my daughter's school, feeling an urgent need to bring

2.2
Remains of WTC façade.
Source: Joe Woolhead.

> her home. I called my son in Pennsylvania to make sure he was okay. I had no sense, no premonition, that within five years' time, I would turn my professional focus to the challenge of envisioning and delivering a memorial museum at the World Trade Center site in New York City tasked with telling the story of the attacks and their aftermath. On that morning, I was a mother and a wife, focused on the safety of my family; an American, grieving for what I intuitively understood would forever represent a rupture in time—knowing that, going forward, there could only now be a "before 9/11" and an "after 9/11."[1]

Lower Manhattan was left desolate for months: The roads were impassable, and the key infrastructure of electricity and telecommunications services was destroyed, requiring urgent restoration, even if temporary.

> It was amazing what we got done in those first 5, 15, 25 days. We first had to run basically large extension cords to get any power to any remaining buildings Downtown. GE (General Electric) built the equipment that we needed for the two substations at 7 WTC in 3 months instead of the usual two years. We got them open

> for the summer of 2004, and the entire deal was done on a handshake. (Eugene McGrath, then Chairman of ConEd, contributing to the T+15 symposium session on "Infrastructure Issues in Rebuilding the World Trade Center and Downtown Manhattan," May 11, 2018)

John Gilbert, Chief Operating Officer (COO) of the Rudin Management Company, led a task force on utilities, where he focused on the commercial users Downtown—the tenants such as the lawyers, investment bankers, and insurance offices—and presented those findings to the utility and service carriers.

> That day, on 9/11, all electricity went down, and all cell towers; only the Blackberries worked until about 5 p.m. that day. What we discovered as a result of the telecommunications collapse on 9/11 was that there was carrier redundancy but not network redundancy. The manhole that blew up held all the carriers. Merrill Lynch had to run their own fiber optic cable across the Hudson and also set up microwave dishes from its tower, providing network backup. Five buildings then provided their rooftops to establish the wireless network for Lower Manhattan. So, the telecommunications solution combined terrestrial connections and wireless connections. (John Gilbert, panelist, T+15 symposium session on "Infrastructure Issues in Rebuilding the World Trade Center and Downtown Manhattan," May 11, 2018)

All major retail stores south of Chambers Street were shut down. Small businesses, whether coffee shops, restaurants, or specialty retail, were shuttered; this also applied in Chinatown and parts of Tribeca to the north of the devastated area. The neighborhood was deserted.

For several months, all pedestrian or vehicular access was blocked below Canal Street, except for emergency runs and the transport of construction workers at the site of "The Pile," as it was then called.

Residents of Battery Park City and other downtown residential buildings had to find alternative accommodation, either in uptown hotels, sublet apartments, or with friends or relatives elsewhere in Manhattan or the metropolitan region. For at least three months, acrid, poisonous smoke hovered over Lower Manhattan and infiltrated the neighborhoods in Brooklyn, Jersey City, and Midtown Manhattan, slowing recovery efforts and always providing a constant reminder of the catastrophe.

Reflections of Natalie R. Cohen[2]

9/11 was a warm, sunny day, and our two boys were attending two different schools: one on the Upper West Side of Manhattan and the other near Waterside in the East 20s. We live(d) on the 12th floor in Battery Park City facing into the World Trade Center site. My husband was in between jobs, so he was able to take our older son to school on the Upper West Side, and I took the younger one that morning. Another mother invited me to walk to our offices together in Midtown. Once we separated and I got closer to my office, I realized something strange was going on downtown. I work in financial services, and we had a television in our trading area. And by the time I got to work, everyone was in the trading area. Everyone was crowding into the trading room, looking at the television, and that answered my question. The World Trade Center Towers had just been hit by the two airplanes.

At that moment, I knew life would be different forever. I also knew I couldn't go home that night and wasn't sure of the condition of our apartment or my husband's whereabouts. I called a friend on the Upper West Side to pick up our older son, and I returned to my younger one's school to retrieve him. To my second grader's credit, we walked together from the East 20s to the West 80s, where my friend lived. I thought three of the four of my family were accounted for.

My husband's story was quite different. He reached home after dropping off our older son and saw the first tower come down from our apartment. It blew out our windows, and at that point, all the dust and debris came in, so he, along with other residents, ran down the stairs to the basement. It is hard to imagine how scared they must have been when they heard the second tower come down since they couldn't see what was going on. At one point, building managers told everyone to come out and head to the waterfront, where boats and ferries would take people to New Jersey, Liberty Park. He had run out of our apartment without his phone, so we had no way to communicate. In Liberty Park, he borrowed someone's Blackberry and sent an email that he was alive. We didn't communicate with each other until 10 p.m. that night when he guessed that I might be staying at my good friends' place.

We stayed there for about two weeks. Each day for several days, we went to Battery Park and waited with neighbors until we had permission to visit our apartment. All of Downtown was closed off from 14th Street south. There was an access point, but we had to prove our residency to get through the barrier. I remember noticing the fleet of electric generators that were parked around Lower Manhattan to facilitate Wall Street's trading and clearing activities.

When we were finally granted access to our apartment, we were escorted one by one by the National Guard. We had 20 minutes to gather whatever belongings we could carry on our backs. The apartment was knee-deep in debris from the Trade Center Towers. I remember seeing Korean trading tickets on our living room floor and pieces of glass lodged in the piano and cabinets from the force of the blast. An airplane wing was lodged in the side of the building. Amazingly, my boys' two goldfish, Gold Ranger and Shimmer, were still alive in the apartment without food. (The Petco salesperson had promised me they wouldn't likely live beyond three months ... but I couldn't bear to leave them in the apartment.) I grabbed a Ziploc baggie from under the sink, took whatever water was left in the pipes, and zipped them up for the 12-story hike down the stairwell. They then lived in another friend's salad bowl for many months more.

Our apartment was damaged and uninhabitable, and we didn't have to pay rent until we could return. We moved into a long-term residential hotel and became middle-income homeless people. I remember having a meltdown at The Gap when I went to buy some underwear and socks. People were amazingly helpful—one son's school put together a collection of pots and pans, clothing, and so on for us.

After 11 months, the Ground Zero site was no longer on fire, we moved back. We had taken a family vote, and the kids said, "But that's home." And in fact, when we moved back, the younger (and more dramatic of the two) said, "Home sweet home" as he crossed the threshold.

I'm eternally grateful that no one in our little family was killed or hurt—many other families were far worse off. Local school kids and workers saw people jumping from the buildings. Fortuitously, I had switched jobs in April 2001. Had I stayed with my former employer located on a high floor at 40 Broadway, I would have seen the planes flying in for the attack.

I'm proud to say that we still live in Battery Park City in defiance of the 9/11 tragedy. (That's what we New Yorkers do, right?) We have been thrilled to watch the new buildings come into being, explore the Oculus, Brookfield, Westfield, and Fulton Center, and invite our friends and relatives to enjoy the food, events, parks, and redesigns in our beautiful, walkable, restored neighborhood.

Assessing the Damage

The damage to the WTC and surrounding buildings and to all the infrastructure in Lower Manhattan was catastrophic, necessitating early estimates of the extent of the damage and the probable costs of rebuilding and repair by many stakeholders and experts in NYC.

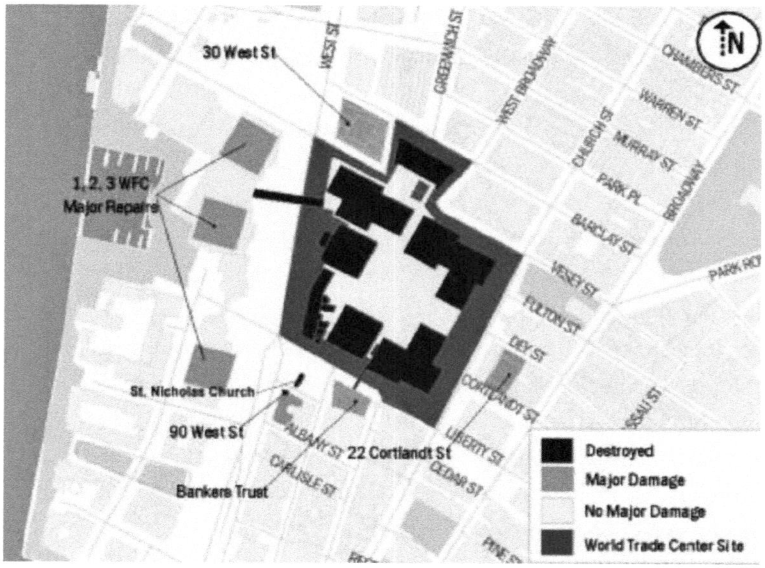

2.3

Map of building damage.

Source: Regional Plan Association.

Reflections of Ric Clark[3]

John Zuccotti, the Chairman of Brookfield USA, stepped into action on September 12 and was quickly appointed [Interim] Chairman of the newly formed Lower Manhattan Development Corporation (LMDC) by Governor Pataki.

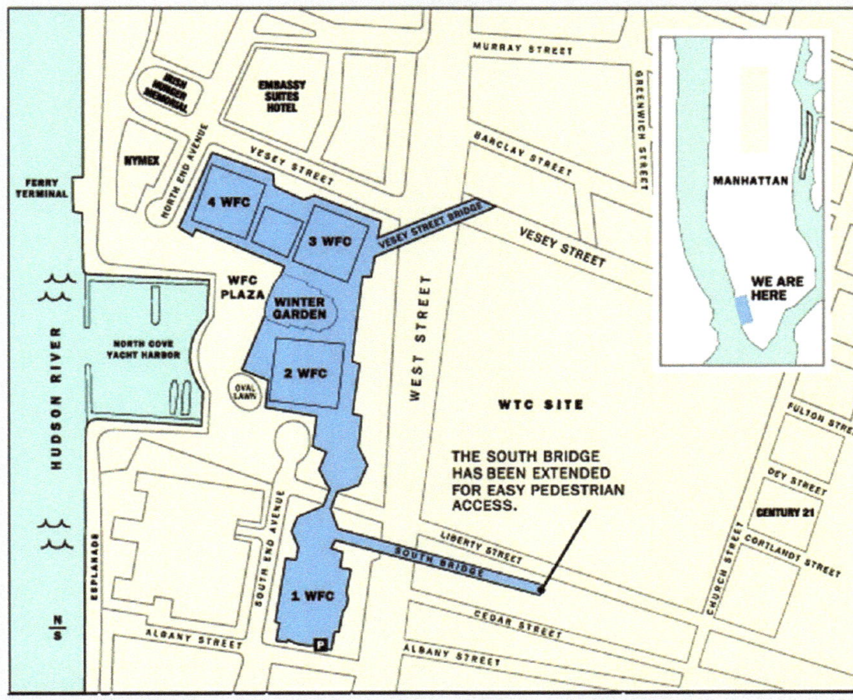

2.4

Layout of Brookfield Place.

Source: Image courtesy of Brookfield Properties.

The damage to the World Financial Center (WFC) by the 9/11 attack was severe. The Winter Garden was completely destroyed, as was the pedestrian connection between Merrill Lynch's occupancy of buildings WFC 2 and WFC 4.

The destruction of the Winter Garden required NYC to decide upon condemnation. This took six months, and none of the retail could be rebuilt during this time. However, the rebuilding of the Winter Garden was completed within one year of the 9/11 destruction.

Both of Merrill Lynch's buildings suffered substantial damage, as did the American Express Building to the north of the Winter Garden. All of the retail establishments were severely damaged and unable to be occupied for years.

We faced new requirements for security at the WFC: The Anti-Terrorism Task Force, in addition to the NYPD [New York Police Department], recommended heightened security measures to the perimeter, including the installation of security

2.5

The damaged Winter Garden.

Source: Image courtesy of Brookfield Properties.

2.6

Rebuilt Winter Garden.

Source: Image courtesy of Brookfield Properties.

bollards, a measure that was at odds with the NYC Planning Department. These were eventually installed at a cost of approximately $1 million.

Part of the problem in our reconstruction and renovation was that FEMA [the Federal Emergency Management Agency] would only pay for the re-institution of what had previously existed. If changes were being planned, then the process became severely protracted. For example, the rebuilding of the underpass from WFC to WTC was planned to be completed in 2009 in conjunction with the building of 1 WTC. But in 2012, there were still disagreements between FEMA, PANYNJ, and NYS: The Port Authority was still resolving the passage of utilities to the WTC site from under West Street, and the NYS Transportation Department was still resolving the design with respect to the restoration of West Street's (aka Route 9A's) road structure.

This became a particular problem since the approaches to the renovated Brookfield Place could not be completed. So Brookfield stepped in and did the work: We used fixed-price contracts, charged a 1 percent management fee, saved $50 million of the originally estimated cost, and completed the West Street frontage work by 2015.

Important Role of the Port Authority

As the developer and owner of the 16-acre WTC complex, the Port Authority was in a unique position regarding the rebuilding of the World Trade Center site, with a multitude of responsibilities and challenges.

First, as the ultimate owner and economic beneficiary of the WTC site, the loss of substantial rental revenue for the Authority was an additional financial blow. Just six weeks before 9/11, Larry Silverstein, owner of Silverstein Properties, had completed negotiations with the Port Authority for the "purchase" of the WTC.

This was, in fact, a 99-year lease of the office buildings, as the Port Authority retained ownership of the ground. Under the terms of the agreement, Silverstein agreed to pay the Port Authority an annual rent of $120 million, maintain the operating condition of the buildings, and also replace or rebuild in the event of any destruction. There was also a lease with Westfield, a large owner of shopping malls, for the 400,000 square feet of WTC retail space.

Adding to this challenge, the Port Authority offices had been located in 1 WTC, and the lives of 78 of its own community had been lost on 9/11. The sudden lack of its own work premises also made it more difficult to attend to the recovery of the infrastructure and operations for which it was responsible.

Second, the Port Authority owned and operated the Port Authority Trans-Hudson (PATH) commuter railroad from New Jersey, which brought approximately 100,000 workers daily through two tunnels under the Hudson River to Manhattan, 60,000 of whom came to the WTC PATH Station. Therefore, much was dependent upon rebuilding the PATH train station at 1 WTC. As Glaeser and Shapiro noted, "The downtown area's continued strength comes primarily from its access to New Jersey (via the PATH trains)."[4]

The transfer of managing the WTC's rental revenues to the private landlords, Silverstein and Westfield, was intended to strengthen the economic support for the important PATH transit system, and now it was their contractual responsibility to rebuild the office tower and retail complex. That rebuilding, however, was to be more complicated and politically charged given the continued land ownership of the Port Authority and the associated commuter system.

Reflections of William Plunkett[5]

The week after 9/11, I joined a meeting convened by Larry Silverstein and a team of lawyers. Silverstein was overwhelmed with passion, conviction, and determination to rebuild, though this resolve was sorely tested over time. I worked as an advisor to Larry for nine years but went out of my way to keep an extremely low profile. We established an effective working relationship with Governor Pataki.

In general, the ugly head of politics reared often. The first three to four years were an endless battle; not a day passed without something causing trouble or someone changing positions. The situation became adversarial, though it should not have been, given the need to rebuild such an important city area. The Press often exacerbated the controversies.

Larry's blind spot might have been in understanding or having patience with the political processes. My job often was to explain the many issues with which the Governor had to deal. Howard Rubenstein helped by handling public relations, while lawyers at Wachtell, Lipton, Rosen & Katz did the day-to-day work of translating the situations into what each meant legally. This was often different from the perceived public impact and response.

Economic Impact of the Destruction

In her detailed discussion of the financial complexity of rebuilding the WTC, Lynne Sagalyn makes an estimation of the damage: "The total cost of the 9/11 terrorist attack—insured losses on NYC, DC, and airlines—was $24.4 billion, by far the costliest event of terrorism."[6]

Focusing on the loss of office buildings, Robert von Ancken, Chairman of Landauer Valuation & Advisory, a division of Newmark Grubb Knight Frank, provided an informative overview at the April 10, 2017, panel discussion on "Envisioning the Office Market Downtown, Post 9/11":

> Before 9/11, Lower Manhattan had approximately 90 million square feet of office space. This was 25 percent of NYC office space. Rents were at $45 per square foot, and vacancy was 7.5 percent.
>
> In 2001, post 9/11, there were 64 million square feet of office space remaining. That is a 20 percent reduction. 13 million square feet at the World Trade Center were destroyed. At least 12 million square feet of adjoining buildings were

damaged. Over 25 million square feet of space was lost. The vacancy rose to around 10.6 percent [of remaining stock]. Asking rents plummeted. (Robert von Ancken, panelist, T+15 symposium session on "Envisioning the Office Market Downtown, Post 9/11," April 10, 2017)

Within three weeks of 9/11, the NY Fiscal Policy Institute issued a preliminary report estimating the total lost output to the NYC economy to be $16.9 billion, with 108,500 jobs, 2.4 percent of the city's total, lost within the first month.[7] The industries most immediately affected were financial services, retail trade, restaurants, air travel, and tourism.

Both the City and State Comptrollers reported in early October regarding the estimated impact on NYC's economy and revenues. The City Comptroller estimated the total economic loss at $20.8 billion, with a risk to city revenues of $738 million in FY 2002 and a further risk of $567 million in FY 2003. The State Comptroller's report raised concerns about the financial and revenue losses to nearby counties and jurisdictions, which, together with the substantial NYC revenue losses, would intensify the economic damage to the State of New York. This report also noted the NYS Governor's request to the federal government for $20 billion in economic aid to NYC, over and above the direct costs of rebuilding the WTC, which were to be covered by insurance proceeds.

Shortly after the attack, a group of consultants and economists ("the consortium") was convened by the New York City Partnership[8] to assess the economic damage and costs to industries and the infrastructure of NYC, with particular emphasis on the economic costs to Lower Manhattan.

Six weeks later, in November 2001, the consortium reported in *Working Together to Accelerate New York's Recovery: Economic Impact Analysis of the September 11th Attack on New York City*, concluding (p. 5) that "[t]he September 11th attack caused massive damage to New York's economy," with an economic impact "likely to total $83 billion," with a net loss in the city's economy of "at least $16 billion," "after estimated payments of insurance claims and federal reimbursement for rescue, cleanup, and infrastructure repair costs." "An estimated 125,000 jobs will be lost in the fourth quarter of 2001 … [and] New York City will still have a net loss of approximately 57,000 jobs attributable to the attack at the end of 2003."

The concentration of damage in Lower Manhattan was noted: "[it] lost 100,000 jobs and almost 30 percent of its office space," and "the financial services, tourism, and retail sectors felt the greatest impact"; small businesses in all sectors "were and will continue to be the hardest hit," "putting at risk many of the 270,000 jobs that are still located in the area south of Chambers Street" (p. 5).

The report urged that Lower Manhattan be quickly restored as a functioning community for workers, small businesses, and residents; that the downtown infrastructure—the power grid, transportation, and telecommunications—be put on a fast-track rebuilding; that federal economic aid be secured to protect the city's status as a world capital of commerce and finance; and that a plan to redevelop Lower Manhattan be produced by mid-2002.

Further Reflections of Carl Weisbrod

The catastrophe of 9/11 was devastating for Lower Manhattan. All telephone service was out, electricity was severely damaged with the destruction of two substations at the World Trade Center, and the National Guard was bivouacked in Battery Park. You essentially needed a passport to get south of Canal Street for weeks on end. It was a very, very difficult, as well as painful environment.

The Downtown Alliance's role immediately after 9/11 was how to get some aid to the small businesses that were in Downtown and how to get some life into Downtown New York.

The residents had largely left—the residential population was understandably cut in half, at least temporarily—because people could not function in Lower Manhattan. Stores were all closed, not just large businesses, but the basic services—where to get a cup of coffee, where to buy food, where to do anything. The amenities that are required for both commercial and residential life were lost. And those businesses, for the most part, small businesses, had nowhere to turn. The government was not geared up yet because, for all the aid that had been promised and ultimately delivered by Washington, which was really astounding and critical to the ultimate redevelopment of Lower Manhattan, that aid didn't start to flow for a long time.

So, what the Downtown Alliance did was organize and generate loans to small businesses. These totaled about $20 million and were sight loans, character loans, essentially soft loans, which could really be characterized as grants because very few of them were ever repaid or were expected to be repaid. We did this so that small businesses could clean up, open, stay open, and provide the basic services that a community needs until it could get stabilized. These loans could also be used to pay their staff who were left suddenly unemployed and without income. We were able to organize the loans quickly and disburse right away to these small businesses.

2.7

Carl Weisbrod walks amongst the devastation of Lower Manhattan.

Source: Alliance for Downtown New York.

Capital Cost of the Damage

At the end of the initial recovery phase, other economic consultants[9] reported on the capital costs for rebuilding:

> The capital losses directly resulting from the attack on the World Trade Center include the damage and destruction of the public and private physical capital stock in Lower Manhattan and the permanent loss of the productivity of the almost 3,000 workers who lost their lives in the attack. The physical losses include about 30 million square feet of commercial office space and about 100 retail stores in the World Trade Center area, subway tunnels (Lines 1 and 9), the Port Authority Trans-Hudson train station at the World Trade Center, the streets surrounding the attack site, and parts of the telecommunications and power infrastructure in Lower Manhattan. In order to put a monetary value on these losses, we first group the components into three categories: 1) the cost of site cleanup and restoration, 2) the cost of replacing the damaged and destroyed buildings and their contents, and 3) the cost of repairing the damage to the infrastructure. Our estimate of the value of each of these losses as of mid-2002 is shown in Table 1 [Table 2.1].
>
> As of June 2002, the site cleanup and restoration were largely completed at a cost of approximately $1.5 billion. The cost of replacing the destroyed buildings in the World Trade Center complex and adjacent area was estimated to be $6.7 billion; the cost of repairing the damaged buildings in the World Trade Center area was estimated to be $4.5 billion, and the cost of replacing the contents was estimated at $5.2 billion. These three categories total $16.4 billion (in $2001). The cost of the repairs to the major elements of the damaged public infrastructure was $3.7 billion.[10]

Table 2.1 Capital Losses[11]

Category	*Estimated Cost ($2001 Billion)*
Cleanup & Site Restoration	1.5
Destroyed Buildings in WTC Complex	6.7
Damaged Buildings in WTC Area	4.5
Contents of Buildings in WTC Complex	5.2
Public Infrastructure 3.7 • Subway 0.85 • PATH 0.55 • Utilities 2.30	3.7
TOTAL (in 2001 dollars)	**21.6**

The destruction of the office buildings meant that many businesses immediately moved from Lower Manhattan to other NYC locations, with the risk that such relocations might be permanent.

> The majority of large displaced tenants signed leases in locations outside of Downtown Manhattan. For transactions larger than 50,000 square feet, 65 percent were signed in Midtown, 17 percent in New Jersey, 5 percent in Westchester County, NY, and Connecticut, and 9 percent in Brooklyn and Queens in New York.[12]

Funds for Rebuilding

Such unprecedented destruction of a major American urban area had not previously been experienced, and there was not an established source of funding for restitution and rebuilding.

Both public and private structures were damaged, so it was anticipated that capital would be sought from public budgets and insurance policies to cover losses and the costs of repair.

In a complicated context, the public-sector operations and physical structures were under a mix of NYS and NYC jurisdictions. Additionally, the defense systems to be put in place against future attacks would be the responsibility of the U.S. federal government—the Department of Homeland Security was established—and the participation of the Department of Defense and the Federal Bureau of Investigations would be vital.

Therefore, the funding would be drawn from all three levels of government, though the rebuilding and operating activities would not be clear-cut, but rather interwoven and multilayered. The severe damage to the economic activities of the city and state, however, precluded the provision of exceptional funding, so the predominant source of capital for rebuilding was to be the federal government.

Federal Funds for Rebuilding the Economy of Lower Manhattan

The economic team of Bram, Haughwout, and Orr of the Federal Reserve Bank of New York described this funding and how it was provided:[13]

> Focusing on this federal aid: the headline amount of federal support dedicated to the city was $20.8 billion. But this amount was not simply made available to the city to use; rather, the purposes and amounts for which the aid was to be used were fairly rigidly defined and to be paid over time as the recovery progressed through different stages.
>
> The broad categories of assistance and their amounts as of June 2002 are outlined in Table 2 [Table 2.2]. Roughly $6 billion was channeled through the Federal Emergency Management Agency (FEMA) for the cleanup of the site and the repair of the damaged public utilities. Restoration of these facilities was largely complete within the first six months following the attack. The remaining funding categories included place-based grants and loans intended to encourage people to live and work in Lower Manhattan and thus hold activity in the area together. The $3.4 billion in grants/loans were direct expenditures of Federal funds that were provided to firms and residents in the area. Longer-term subsidies encouraged businesses to invest in the area and for improvements to the transportation infrastructure, taking people into and out of the area.

Table 2.2 Estimated Federal Support for Cleanup, Rebuilding, and Economic Revitalization in New York City Program[14]

Category	*Estimated Assistance ($2001 Billion)*
FEMA Emergency Response • Cleanup • Public utilities repair	6.0
Grants/Loans • Job creation grants/loans • Business recovery grants • Small business attraction and retention • Cultural attractions funding	3.4
Economic Stimulus • Tax-exempt construction bonds	5.0
Transportation Projects • Intermodal hub • PATH terminal	5.0
Other	1.4
TOTAL	**20.8**

Alternatives to Federal Funding of the Rebuilding

Not all citizens, and particularly urban scholars, supported the contribution of federal funds to the rebuilding of the physical fabric of Lower Manhattan. Glaser and Shapiro were particularly skeptical about national funding of the rebuilding:

> The case for subsidies hinges on the view that there are remarkable intellectual spillovers associated with the downtown area. Indeed, the continuing vitality of the area pre-9/11 gives credence to this view. However, midtown also has dramatic intellectual spillovers. If the activity moves from the dense downtown to the almost equally dense midtown, it seems unlikely that there will be a loss in New York's fertile financial industry. To the extent that employment moves to medium-density areas in New Jersey, there is more of a concern. Nonetheless, the fact that most displaced firms have moved to midtown and the absence of strong evidence for a loss of spillovers due to firms leaving Manhattan makes the case for massive subsidies questionable at best.
>
> We think that it is foolish to engage in a large federally subsidized rebuilding program Downtown. It is surely more efficient to let businesses relocate either to midtown, where there is still plenty of room to build, or to office parks in suburban New Jersey. Downtown is far from the population centers of New York and is built around a port that no longer exists. If firms want to rebuild there, then so be it. The private market shouldn't be blocked, but the future of New York City doesn't depend on building Downtown, and there is no economic reason to explicitly subsidize that activity.[15]

Notes

1 From 2006–2016, Alice M. Greenwald was Executive Vice President for Exhibitions, Collections, and Education and Director of the National September 11 Memorial & Museum. From January 2017 to September 2022, she served as its President and CEO.

2 Natalie R. Cohen is founder of National Municipal Research, an independent consulting and research company focused on U.S., state, and local government finance. She has been a resident of Battery Park City since 1992.

3 From 1984–2020, Ric Clark was with Brookfield Asset Management and its affiliates in various senior roles, including those of Chairman and Chief Executive Officer of Brookfield Property Group.

4 Edward L. Glaeser & Jesse M. Shapiro, "Cities and Warfare: The Impact of Terrorism on Urban Form," *Journal of Urban Economics* 51 (2002), pp. 220–221.

5 William Plunkett is a Senior Counsel of the law firm, Plunkett PLLC. He has provided counsel on major commercial and residential real estate and economic development projects in New York, including several billion-dollar projects such as the post-9/11 World Trade Center redevelopment.

6 Lynne B. Sagalyn, *Power at Ground Zero: Politics, Money and the Remaking of Lower Manhattan* (Oxford University Press, 2016), p. 382.

7 *Economic Impact of the September 11 World Trade Center Attack*, Fiscal Policy Institute, September 28, 2001.

8 The New York City Partnership, *Working Together to Accelerate New York's Recovery* (2001), p. 5. The consortium included AT Kearney, Bain & Company, Booz/Allen/Hamilton, The Boston Consulting Group, Harris Interactive, KPMG, McKinsey & Company, and PwC Consulting, as well as consultation with almost 400 specialists and close to 25 companies, associations, or educational institutions.

9 Jason Bram, Andrew Haughwout & James Orr, "Further Observations on the Economic Effects on New York City of the Attack on the World Trade Center," *Peace Economics, Peace Science and Public Policy* 15(2) (2009), Article 2, pp. 2–3. DOI: 10.2202/1554-8597.1159.

10 Bram, Haughwout & Orr, "Further Observations on the Economic Effects on New York City" (2009), Note 5 states:

> These damage estimates are somewhat lower than estimates reported immediately following the attack. The main source of the difference was the estimated cost of cleaning up and restoring the World Trade Center site. Estimates by both the New York City Office of the Comptroller (2001) and New York City Partnership and Chamber of Commerce (2001) turned out to be roughly $3–4 billion higher than the actual cost. Part of the reason for this higher estimate was the inclusion of costs related to repairing potentially hidden or less-visible structural damage to the site. A report by the Government Accountability Office (2002) compares these initial damage assessment estimates.

11 Derived from Bram, Haughwout & Orr, "Further Observations on the Economic Effects on New York City" (2009), p. 3.

12 J. J. Studley,*Studley Report: Special Edition* (2002, September 11), p. 2.

13 Bram, Haughwout & Orr, "Further Observations on the Economic Effects on New York City" (2009), pp. 4–5.

14 Derived from Bram, Haughwout & Orr, "Further Observations on the Economic Effects on New York City" (2009), Table 2, p. 5.

15 Edward L. Glaeser & Jesse M. Shapiro, "Cities and Warfare: The Impact of Terrorism on Urban Form," *Journal of Urban Economics* 51(2) (March 2002), p. 222.

Chapter 3: The Years of Intense (and Often Controversial) Planning to Rebuild

In *Up from Zero,* his detailed account of the rebuilding process in the immediate years following the 9/11 catastrophe, Paul Goldberger observed:

> There was not only no precedent for dealing with the enormity of the loss; there was no system for figuring out what should happen next. Should the towers be put back? Should they be replaced by a triumphant substitute or by a sober memorial? Or both ... Who should be empowered to decide all of this, and by what means? And who should pay for it? Although it took less than two hours for the towers of the World Trade Center to collapse, it has taken more than two years to figure out these questions, and some of them still have no answers.

And he further noted: "Planning is not always a particularly rational process in New York City under the best of circumstances, and these circumstances were stunningly different from anything anyone had ever experienced."[1]

Throughout his book, Goldberger profiled the many political and civic leaders who staked a claim to the type of rebuilding, and also the leaders of important organizations and businesses who had constituents who were directly impacted and tragically suffering. He included the various academic and professional groups that came together to develop and share their ideas of what should be built and suggest processes to guide that rebuilding. In all, just the planning of the restoration of the area was to be a process fraught with political conflicts, vocal citizenry, and diverse frameworks of reference.

Reflections of T. J. Gottesdiener[2]

Perhaps there should have been a preliminary formulation of urban principles for such an important site before the physical planning was first undertaken. Certainly, one important objective was the connection with the urban context at the street level, a principle promoted by the City Planning Commissioner, Amanda Burden, in response to the first iteration of plans (by Beyer Blinder Belle), which had the replacement of the new WTC towers on a platform. This sent the wrong message, instead of the new plan knitting the city together.

DOI: 10.1201/9781003476139-3

However, there was some significant commonality of aspects in the planning process between the various stakeholders and the early teams of architects: that there should be open public space, that the street grid should be extended through the WTC site, that the existing topographical contours were to be utilized, that the footprints of the original towers were to be respected, and that density was necessary. The Port Authority was focused on utilizing some of the $21 billion federal grant to build a transit hub similar to such structures in London and Germany.

The contentious issues were: the distribution and arrangement of the density, the form of the memorial, the commercial use of the site, and other items representing fractious political agendas.

Rebuilding Envisioned

While the federal government seemed to have the broadest responsibility for funding the rebuilding, the NYS Governor's Office proposed a plan to contain the diverse resources needed for the rebuilding effort. In late November 2001, Governor George Pataki announced the formation of the Lower Manhattan Development Corporation (LMDC) as a joint NYS and NYC entity dedicated to overseeing the rebuilding and economic revitalization of Lower Manhattan.

The Lower Manhattan Development Corporation

With this new joint state–city responsibility for planning and coordinating the rebuilding of the WTC site and the state-authorized involvement in allocating federal funding, the LMDC held the most prominent position with respect to decisions to be made about the functional and physical form that would emerge. Its immediate challenge was to work with the myriad of distinctly different stakeholders. The newly appointed Chairman, John C. Whitehead, formerly of Goldman Sachs and a one-time Undersecretary of the U.S. Department of State under President Reagan, appreciated that this required a significant period and process of listening. Various advisory councils were established to address some of the most clearly defined concerns, and the organization set to work to be responsive across an extraordinary range of issues.

The concerns of the victims' families for an appropriate memorial to their lost ones and their need for financial and emotional support varied significantly from the economic interests of major firms wishing to re-establish their operations. By being an authoritative focal point, even though often drawn in many directions, the LMDC was able to commence planning for the rebuilding for functional recovery of Lower Manhattan while simultaneously managing the civic distress.

In filling out the executive ranks, Louis R. Tomson, who had been first Deputy Secretary to Governor Pataki, was appointed the first Executive Director and President of LMDC at its formation, and Kevin Rampe, first Deputy Superintendent and COO of the New York State Insurance Department became Executive Vice President. In February 2002, the urban planning practitioner and scholar Alexander Garvin was appointed as Vice President for Planning, Design, and Development.

Planning Blueprint

Early in 2002, the LMDC commenced its critical planning activities. Alex Garvin remembers:[3]

> In April 2002, the LMDC issued a draft of its Principles for Action, which called for an open decision-making process, rapid and long-term development, and comprehensive, exceptional, and diverse planning. At the same time, it released a draft of its Preliminary Blueprint for the Future of Lower Manhattan, which described the requirements for future public and private development in Lower Manhattan, including a memorial, the restoration of all or part of the original street grid, pedestrian access improvements across West Street, a coordinated mass transit plan with a distinctive hub, and the creation of a mixed-use, 24/7 community with exceptional public spaces.
>
> The LMDC held a public hearing on those documents at Pace University in May (2002) and, based on public comment, made minor revisions and formally adopted them in July (2002).

The Preliminary Blueprint comprised general principles and guidelines incorporating a variety of suggestions from many of the self-formed advisory professional groups such as New York New Visions. LMDC recommended:

- Extensive improvements in infrastructure, including an intermodal transit hub connecting the PATH train and the area subways.
- The creation of a park inclusive of some form of memorial.
- A museum devoted to the 9/11 attack.
- The provision of space for cultural activities.
- The reconstruction of the original street grid through the WTC site.
- Continuation of the Port Authority's commercial program with respect to the ground leases to Silverstein and Westfield.

To accomplish these goals, the report identified three possible scenarios:

Scenario 1: To ensure that Lower Manhattan resumes its role as "NYC's Second Central Business District" by emphasizing the construction of 10 to 12 million square feet of new commercial office space that would yield 60,000 new office jobs by 2022, 5,000 units of residential development, and 2 million square feet of retail development.

Scenario 2: To emphasize residential and retail development and generate a new center for tourism and the arts. This would entail producing 17,000 units of housing for approximately 37,000 additional residents and 2.5 million square feet of retail space. New office construction of 5 to 6 million square feet would be encouraged, yielding 30,000 new office jobs, half of that anticipated in Scenario 1.

Scenario 3: For Lower Manhattan to emerge as a "Global Downtown," rebuilding would comprise substantial office construction of 15 to 18 million square feet

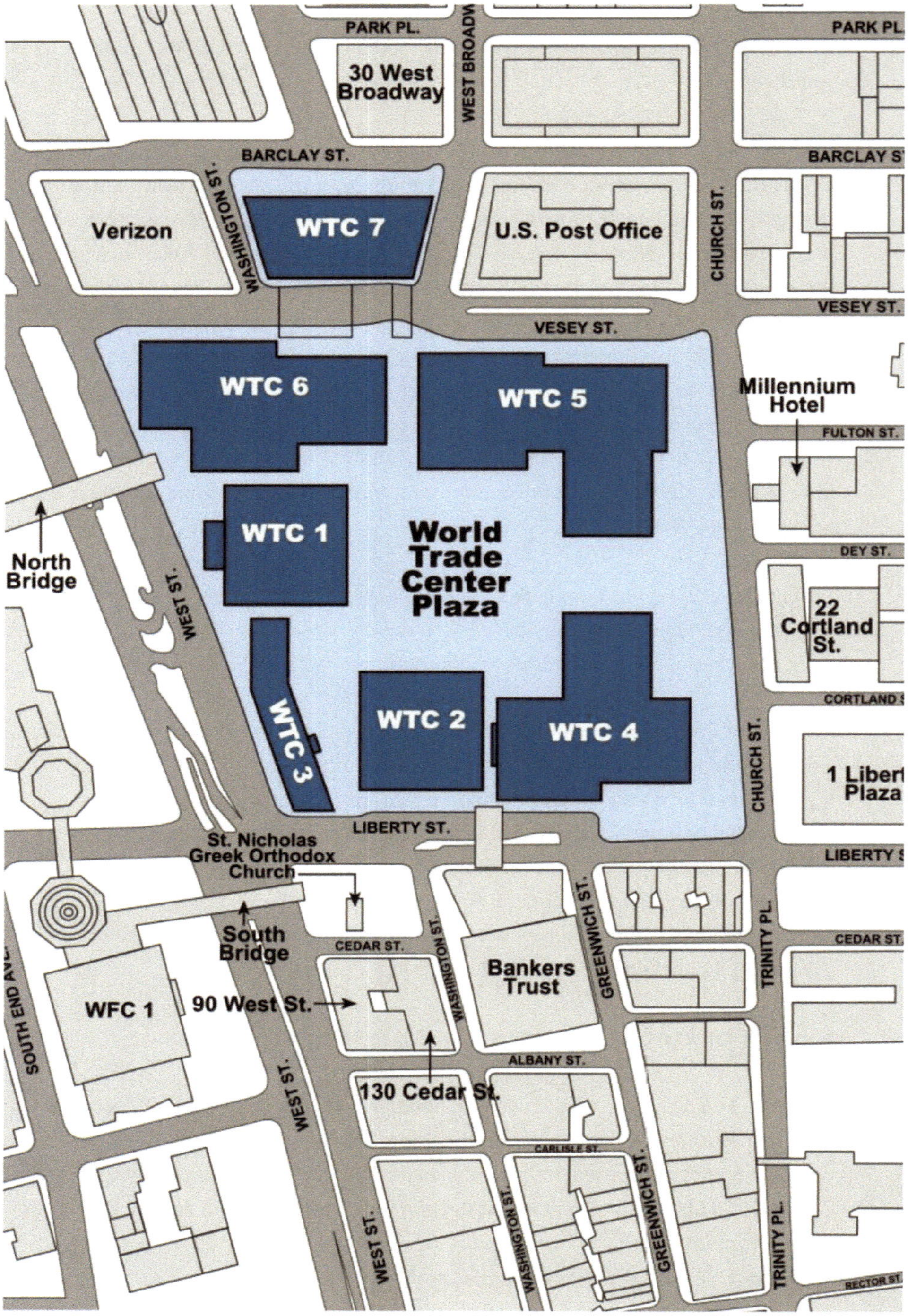

3.1

Early site planning exercise in distributing office buildings.

Source: JDA/Silverstein Properties.

to yield space for 90,000 new office workers. Retail development would still be encouraged at 2.5 million square feet, but with 7,000 units targeted for residential development, population growth would be less than that envisioned for Scenario 2 but greater than for Scenario 1.

The report provided ample detail of the range of services and amenities, including the transportation infrastructure needed to bring any of the three scenarios to fruition.[4]

Alex Garvin further describes[5] the benefit of this achievement:

> It [LMDC] set forth a reconstruction program that included expanding the residential population, making lower Manhattan a "vibrant place with daytime and nighttime activity," investing in better subway and train services, and creating a "comprehensive park and open space system."

Assessment of the Future Potential

In September 2002, a team of economists, planners, and real estate specialists presented their report, prepared for the Civic Alliance,[6] to assess the future potential for Lower Manhattan and develop recommendations to strengthen its role in the economy of the entire city and metropolitan region, asking:

- Which industries and activities will shape the economy of Lower Manhattan in the coming decades?
- How should dollars, land, and other resources be allocated among commercial, residential, cultural, and other uses while assisting workers and families who were affected by the attacks?
- How should a revitalized Lower Manhattan relate to the economy in the city and other parts of the metropolitan region?

The study team found that several key areas were fundamental to the recovery of the Lower Manhattan economy:

- rebuilding and enhancing its infrastructure and amenities;
- encouraging continued residential development to enhance the 24-hour character;
- strengthening and integrating the Lower Manhattan economy as a global center, regional hub, and diverse community;
- developing a citywide and regional economic development strategy to support broad industrial sectors to generate a dynamic, diverse, and equitable economy.

Real estate economist Hugh Kelly noted that Lower Manhattan and NYC had rebounded from severe downturns in the 1970s and early 1990s and that almost 80 percent of the office employment displaced by 9/11 had remained within the region's urban hub, both factors offering the promise that the financial sector and its important clustering of key functions could again return in strength to Lower Manhattan.

Urban specialist Richard Florida noted that Lower Manhattan could attract important creative industries, including new technology, professional services, education, and the arts, if substantial investments were made to transit and other infrastructure and by providing substantial residential space.

The report portrayed two alternative visions for Lower Manhattan's economic future:

Scenario 1: A *Rejuvenated Global Office Center* with a rebuilding of the destroyed 13 million square feet of Class A office space to accommodate as many office jobs as possible, suggesting a reconcentration of securities and related financial and professional services. There would be expanded retail and restaurant amenities to cater to office workers and the assumption of some increase in the residential population. Substantial investments in the Downtown transportation system would be necessary to make the area more accessible to commuters. Total employment would grow by 115,000, and the annual number of visitors would increase to 10.8 million by 2010.

- The risks of this strategy were: It would include competition from new office construction in Midtown and Northern New Jersey that could leave an oversupply of space in Downtown; it would depend on a fast pace of economic growth in the nation as well as the region; and it could leave the Downtown economy vulnerable to the boom-and-bust cycles in the financial industry.

Scenario 2: A *Transformed "Creative Hub"* with a more diverse mix of industries and less emphasis on regrowth in the financial sector and substantial growth in the number of residents, cultural activities, public amenities, and visitors. Under this scenario, all damaged office space would be repaired, and new office construction would be restricted to 2 to 5 million square feet by 2010; total employment would grow by 65,000 in that timeframe, knowledge-based industries would increase from 21 percent to 24 percent of total employment while the financial services' share would decline to 22 percent from 30 percent, and the number of annual visitors to Downtown would double from 7.2 million to 14.4 million by 2010.

- The risks of this scenario were: Without a more substantial build of new office space, it would risk losing important sectors of the financial services and other key export sectors; the more diverse industries could fail to provide adequate career ladders for New York youth. In addition, success would depend on attracting sufficient new residents and cultural activities to fully evolve into a 24-hour community.

On December 12, 2002, Mayor Michael Bloomberg unveiled his vision for Lower Manhattan, recommending a series of plans and investments to reinvigorate the area outside of the WTC site with amenities such as parks and schools to bolster additional housing as well as some new office buildings. His vision was to line the streets with trees and begin planning for a more extensive use of the waterfront, particularly along the East River.

As commented by the editors of *The New York Times* on December 13, 2002:

> Mr. Bloomberg wants to turn the intimidating neighborhood of Wall Street canyons into a place that is both elegant and down to earth. It is a Lower Manhattan that invites people to come—a perfect response to the terrorist attack that tried to scare humanity away.

Notes

1 Paul Goldberger, *Up from Zero*(Random House, 2004), Prologue, pp. xi–xii.

2 T. J. Gottesdiener is an architect and was Managing Partner of the New York office of Skidmore, Owings & Merrill (SOM). In 2023, he joined The Georgetown Company to work as an architect-developer.

3 Alex Garvin, panelist, at T+15 Opening Symposium session, December 8, 2016.

4 Lower Manhattan Development Corporation, *Revitalization Strategy for Lower Manhattan*, Working Paper (December 2002).

5 Alex Garvin, panelist, at T+15 Opening Symposium session, December 8, 2016.

6 *Economic Development Working Group Reports*. Prepared for the Civic Alliance to Rebuild Downtown New York, Alice Rivlin and Rosemary Scanlon, Co-Chairs, September 2002.

Chapter 4: Rebuilding Commences

While the planning for the WTC site continued to be resolved, there were imperatives for rebuilding in the surrounding area that were not fully contingent on the specifics of that site. These included the rehabilitation of many buildings nearby that were damaged to varying extents and the infrastructure of streets, utilities, and transport to, from, and through the area.

7 World Trade Center

There was also a building that was not technically within the 16 acres of the WTC but rather was just across Vesey Street to the north. This was known as 7 World Trade Center (7 WTC) and had been built by Larry Silverstein in 1987. Although not directly targeted by a plane, the collapse of the Twin Towers damaged 7 WTC beyond rehabilitation.

The feasibility of rebuilding 7 WTC was aided by the comparable discount of Lower Manhattan rents versus those of Midtown Manhattan, and critically important was the tax incentive provided by NYS, which was valued at approximately $28.5 million.[1] If there were questions about what would be done with the WTC site, with Larry Silverstein's actions, there was no doubt about what would happen at 7 WTC—he was going to rebuild and as soon as possible.

Further Reflections of T. J. Gottesheimer

Meanwhile, within a week of the devastation of 9/11, Larry Silverstein announced that he would commence the rebuilding of 7 WTC, and SOM was brought on to design the new building. Yet, the planning of this building was connected with the planning for the 16-acre WTC site, particularly the concern over the connections or view corridors through to West Street that were proposed for Vesey Street, and the planning by the Port Authority for the reconstruction of the PATH train terminus with the view corridor of Greenwich Street, which was to extend through the 16 acres. Silverstein had proposed rebuilding the 2.1 million square feet of 7 WTC, but this could not be accommodated on the modified site—7 WTC needed to be set back from Greenwich Street. John Whitehead, Chairman of the LMDC, played an important role in pushing for a solution.

DOI: 10.1201/9781003476139-4

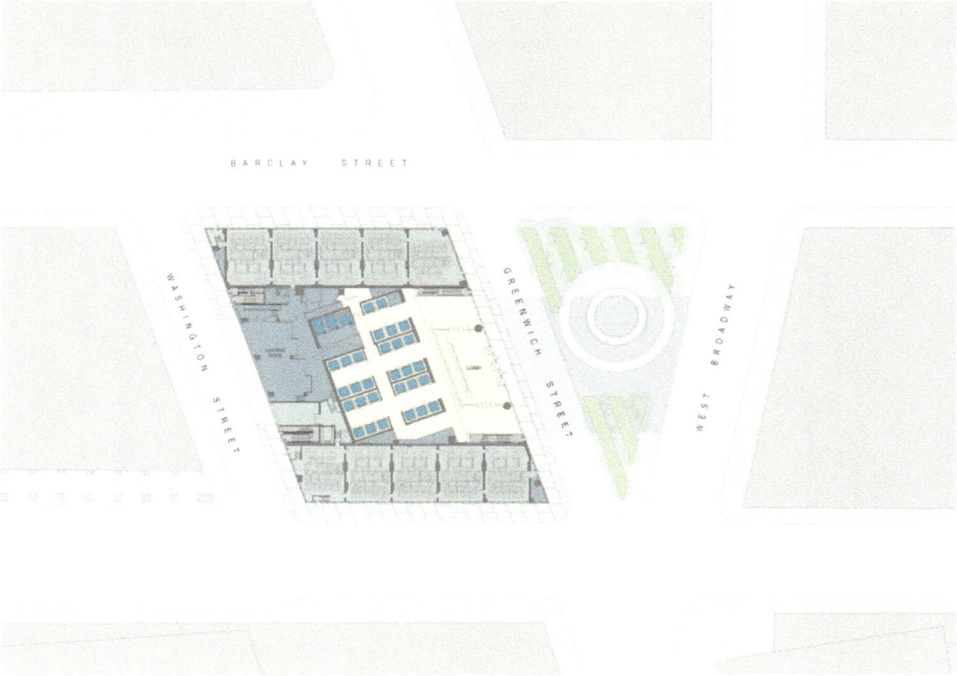

4.1
Ground site plan of 7 WTC.
Source: Silverstein Properties.

4.2
David Childs of Skidmore, Owings & Merrill.
Source: Skidmore, Owings & Merrill.

4.3
Completed 7 WTC.
Source: Ruggero Vanni ESTO/Silverstein Properties.

Private Rebuilding Intersects with Public Infrastructure

The proximity of 7 WTC to the main WTC site meant there were significant interconnections:

1. The 7 WTC design would form a visual and functional composition with the rebuilding of the 16-acre WTC plot.
2. Between the buildings were city streets and utility lines that would need to be shared.

Awkwardly overlapping the boundary of 7 WTC and the public street space between the sites was the main electrical substation for western Lower Manhattan, which also had been destroyed when the buildings collapsed. This was possibly the first clearly defined mutual dependency between the private and public sectors to be formally addressed. Resolving it laid bare the complexity—physical, economic, administrative, and functional—for the rebuilding effort.

The LMDC Chief Planner, Alexander Garvin, in discussion with the authors, recounts:

> The most pressing reconstruction project facing Lower Manhattan was rebuilding the two electric power substations that had occupied the bottom

160 feet of Seven World Trade Center, a building owned by Larry Silverstein, built on property that had been leased to him by the Port Authority.

Speed was of the essence. That meant proceeding without public review and certainly avoiding the time-consuming Uniform Land Use Review Procedure (ULURP). Consequently, in my first two weeks as vice president of the Lower Manhattan Development Corporation (LMDC) in charge of planning, design, and reconstruction, I scheduled a meeting with Larry Silverstein, the building's owner, to decide on the best and most expeditious method for reconstruction.

I was determined to make the rebuilt Seven World Trade Center an improvement on its predecessor. The destroyed building had created a major problem from the moment it had gone into construction. It severed the connection between the Financial District and Tribeca by occupying one block of Greenwich Street, which, along with the rest of the World Trade Center, eliminated north–south traffic between Church and West Streets. I believed that the restoration of traffic flow on Greenwich Street was essential to a thriving Lower Manhattan.

When I met with Silverstein, whom I had never met, his deputy, Jeff Wharton, and his architect, David Childs, a classmate of mine from the Yale School of Architecture, I discovered that they also wanted to restore through traffic to Greenwich Street. Their first proposal was to cut out 60 feet at the bottom of the eastern end of Seven WTC. When I told them that was not acceptable, they displayed an alternative that was exactly what was eventually built. I told them that was perfect. "Unfortunately," they replied, "that is not feasible because it requires 26 feet of Vesey Street, which is owned by the City of New York."

They believed that strained relations between the City and the Port Authority made the transfer unlikely and that it would take at least 18 months of controversial ULURP public hearings to be approved—an unconscionable delay. Moreover, the Port Authority was not sure where its property started, and Vesey Street ended because its records had been destroyed on September 11.

Despite years of experience with the city procedures, I did not have any idea how to complete this transaction. Nevertheless, I told them I would find a way to transfer the property if they invited Deputy Mayor Daniel Doctoroff and me to a meeting [that] they were planning to have with the Port Authority and ConEd to discuss this. They were both skeptical. But Childs, who knew me well, believed that if a sufficiently speedy transfer were possible, I would make that happen.

Doctoroff was my friend and my client for the plans I created to bring the 2008 and then [the] 2012 Olympic Games to NYC. We had a very special relationship and trusted each other completely. When I returned to my office,

I telephoned him, told him about the meeting, told him he would be invited to the next meeting, asked him to appoint David Karnovsky, Counsel to the NYC Planning Commission, as his representative, and hung up. That was the first ever of two peremptory phone calls like that that we ever had.

Karnovsky explained to me that if all the parties agreed, Empire State Development Corporation (ESDC) could acquire the property and transfer it to the Port Authority without going through ULURP [Uniform Land Use Review Procedure].[2] We proposed that at the meeting. When it was over, Karnovsky told me that the transaction would take months.

"What is the problem?" I asked. "The City's Law Department has other priorities," he replied. When I returned to my office, I made my second abrupt call to Deputy Mayor Doctoroff, instructing him to inform the Law Dept. that the transfer of 26 feet of Vesey Street to the Port Authority was a city priority and hung up. He trusted me completely and made Vesey Street a city priority.

On April 18, the board of directors of the ESDC approved the General Project Plan that called for the condemnation of 26 feet of property along Vesey Street without any public review. Ownership was transferred in June. All the parties involved were both delighted and astonished that it had taken only a few weeks.

The groundbreaking officially began on November 20, 2002.

4.4
Larry Silverstein opens 7 WTC in front of Jeff Koons's *Balloons* Sculpture. L. to R.: Jeff Koons, Larry Silverstein, Dan Tishman (Construction), Ken Ringler (PANYNJ Director), David Childs (Architect).
Source: Joe Woolhead, courtesy of Silverstein Properties.

As T. J. Gottesheimer Continues

Key thresholds in the development of the "Freedom Tower" were the announcement in November 2003 that the height of the tower was to be set at 1,776 feet and that this tower would be the vertical culmination of the rebuilt WTC site. In December 2003, Governor Pataki held a press conference announcing the building and proposed a completion date of September 11, 2007. The first comprehensive design by SOM was in June 2004.

However, there was significant contention between the Port Authority and the New York City Police Department (NYPD) in terms of jurisdiction regarding security. The NYPD considered height to be the greatest risk, but also insisted that the walls of the lower floors of 1 WTC should be reinforced and much thicker, and that the expected location of the new tower be moved back from West Street and from Vesey Street. The lead on resolving this conflict was taken by John Cahill, Secretary to Governor Pataki, and Stefan Pryor, then President of LMDC. As a result, the walls at the ground level of 1 WTC are made of special explosion-resistant glass, have [a] two-feet thickness on a steel-based frame, and the resolution on the height of the building would be 60–70 floors topped by a spire to achieve the goal of 1,776 feet in height.

Notes

1 CBRE CB/Richard Ellis, *Trade Center Feasibility Study*, October 21, 2005.

2 The process would be initiated when ESDC issued both a General Project Plan for the project in accordance with the Urban Development Corporation (UDC) Act and a New York State Environmental Quality Review Act (SEQRA) "negative declaration." The NYC government would then review the General Project Plan in accordance with provisions of the UDC Act, which require its local government advisory input in connection with any General Project Plan that involves an override of local regulations. Thereafter, ESDC would proceed with condemnation, uncontested by the City of New York. When the transaction was completed, ESDC would transfer title to the Port Authority, which would, in turn, lease it to Silverstein Properties for inclusion in 7 WTC.

Chapter 5: Planning Groups and the Public Interest

In addition to its responsibility for the strategic planning for use of the $21 billion in recovery aid distributed through various agencies by the federal government, the LMDC was also a specific recipient, custodian, and disburser of $2.5 billion for operational and administrative costs in overseeing the revitalization of Lower Manhattan. Therefore, it was well positioned to conceptualize the early plans and designs for what was now referred to as "Ground Zero." The LMDC staff undertook internal site analyses, established the planning blueprint, sponsored design competitions for a Master Plan and key office buildings, and initiated design competitions for a memorial, the new PATH terminal, and a 9/11 Memorial & Museum.

Together with the site's owner, the Port Authority, LMDC issued a request for proposals from architects for the site's Master Plan. Responses were received from 15 architectural firms, and after internal assessments, the architectural firm Beyer Blinder Belle, partnered with the infrastructure engineering firm Parsons Brinckerhoff, was selected. The release of the winning Master Plan was met with mixed critical reviews but a surprising predominance of public disdain. The time allotted for these initial studies had been short, and although the proposal was intended to be a feasibility study incorporating schematic physical planning of the site, the plans that were produced were perceived, particularly by the public and the media, as full architectural resolutions of the built form to be erected on the site.

Since the tragedy of 9/11 had affected so many people in various ways, many other groups and communities formed to consider various aspects of the recovery. One important group to emerge was the Civic Alliance to Rebuild Downtown New York (Civic Alliance), led by Robert ("Bob") Yaro, who was President of the Regional Plan Association in New York. The Civic Alliance was to include the New York Chapter of the American Institute of Architects (AIA), the NYC Municipal Art Society, and representatives from several regional universities. The Civic Alliance, along with the LMDC, Community Board 1,[1] and the Downtown Alliance, held the first hearing for an audience of the public in January 2002, attended primarily by residents of Lower Manhattan. This meeting produced a laundry list of public requests, such as more green space and more residential amenities.

DOI: 10.1201/9781003476139-5

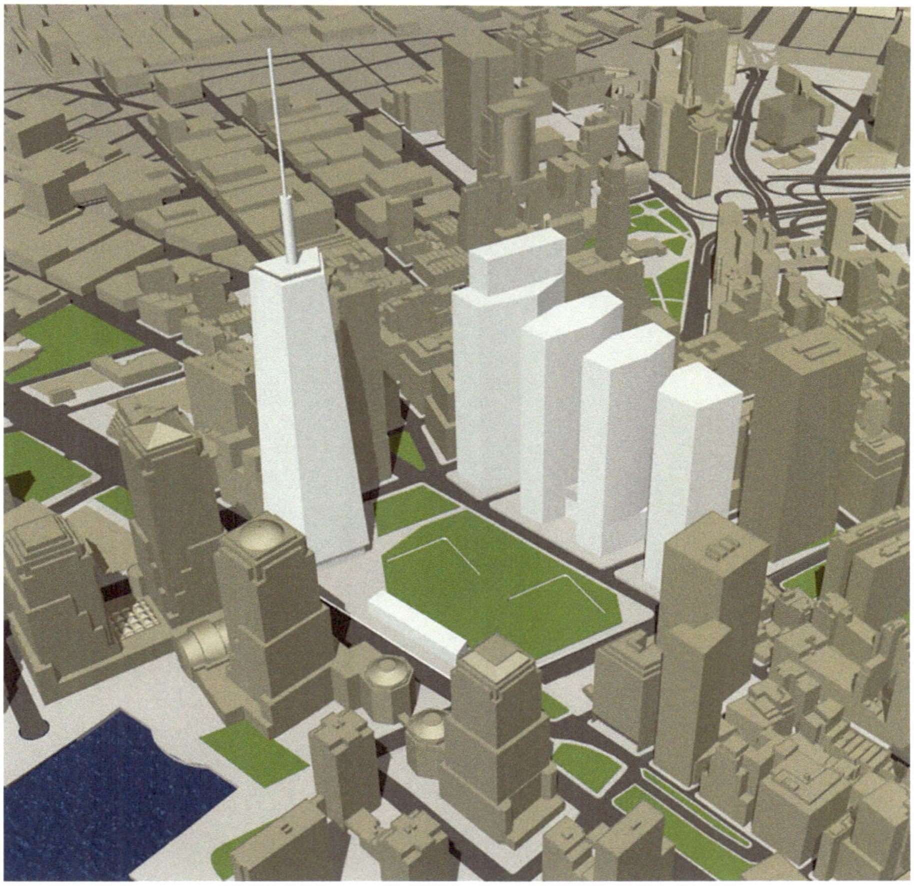

5.1
Beyer Blinder Belle Memorial Plaza concept plan (one of six schemes).
Source: Silverstein Properties.

As the LMDC's conceptual plans were on view during the spring of 2002, Bob Yaro of the Civic Alliance, with the LMDC and the Port Authority as co-hosts, organized a second gathering at the South Street Seaport under the concept of "Listening to the City" to elicit further ideas on the redevelopment of the 16-acre site.

From this event, it was proposed that a larger public forum be held at the Javits Center on a Saturday in late July 2002. At that event, some 5,000 participants were led through a series of electronic votes to comment on and react to the six recently unveiled proposed scenarios of the Master Plan produced by Beyer Blinder Belle, in addition to some supporting explanatory plans produced by Steven Peterson and Barbara Littenberg as Planning Consultants to LMDC.

The public attendees, moderated by Carolyn Lukensmeyer of AmericaSpeaks, discussed, debated, and commented on, sometimes heatedly, the proposals, their objectives, and their architectural forms.

Paul Goldberger summarized the dynamics of the day:

> By the end of the day, the public had made it clear that it found the six plans to be banal, lacking in vision, excessively devoted to commercial concerns, and inadequate as settings for the kind of memorial they had hoped to see.[2]

5.2
"Listening to the City" public meeting, February 7, 2002, at South Street Seaport.
Source: Courtesy of Regional Plan Association.

Tom Wright, then Deputy to Bob Yaro,[3] Reflects

Bob's plan for public involvement in the decisions to rebuild began on the morning after the attack when he met with Stephen Heintz, President of the Rockefeller Brothers Fund. They agreed that "we have to plan, and we need to get all of the major civic groups together to think about the future of the WTC site." On the Friday after the attack, September 15, he attended a meeting with Lee Sander, the former DOT [Department of Transportation] Commissioner for NYC, and Carl Weisbrod, who ran the Alliance for Downtown New York. They agreed on the need for a public discussion, and out of that came the organization of the Civic Alliance to Rebuild Downtown New York. The first Listening to the City event was a smaller forum to test the concept of an electronic town hall forum. It was held at Pace University in early 2002. The next larger events were held at the Javits Center on July 20 and 22, 2002. These forums attracted over 5,000 people.

In setting up these events, Bob felt strongly that he was helping to protect Downtown Manhattan. "We were in danger of losing the third biggest CBD in the US. It was critical to keep a critical mass of commercial office space, and, in those early days and weeks after 9/11, there was a great crisis of confidence." So, with Listening to the City, he wanted to "engage all that angst and energy."

I began working with Bob Yaro at RPA on September 13, 2001, so I was thrown right into this effort from the beginning. In those early months after 9/11, there was an enormous amount of tension and competing interests around the site. This tension radiated from different local communities to a much wider citywide, regional, national, and even global community. And so, understanding those competing demands was really important. I think the civic community writ large had a couple of powerful concerns:

One was that they would be completely ignored. That the power brokers—the major business interests and the public sector—would convene and decide what to do and that there wouldn't be any ability for the civic community, local residents, or other constituencies to be involved. The civic community wanted to be at the table. We had early financial support from the Rockefeller Brothers Fund and the J. M. Kaplan Fund. Later, we were generously supported by the law firm Shearman & Sterling to open up the planning process to a wider range of participants.

There was also a worry that the public urge for speed and haste was going to embed poor decisions that we would have to live with for generations. When we held the RPA Regional Assembly in the spring of 2002 to invite all the groups to the table and coordinate activities, I vividly remember Paul Goldberger, as our keynote speaker, saying, "I worry about the urge for speed right now. And if we rush and get it wrong, we're going have to live with it for a long time." At the time, just about every editorial page in the city was pushing for decisions to be made. Paul's argument was important—that this should take some time. I'm glad that we were able to push back against the rush to do something too quickly.

To open up the process and allow for more deliberation, we convened dozens of groups, including The New School, the AIA, and the Municipal Arts Society, to form the Civic Alliance to Rebuild Downtown New York. Many groups were already launching their own programs, such as New York New Visions and Imagine New York, and each was addressing important questions. Unlike some other efforts, which were pushing for central control, the Civic Alliance provided an opportunity for voluntary collaboration and coordination. We looked at a range of topics: How should the streets of Lower Manhattan be restored? What mix of development and open space should be achieved? And, of course, how can rebuilding the economy and community of Lower Manhattan coexist with a memorial to the tragedy of 9/11? Different working groups were established within the Civic Alliance and set out to answer these questions and share information. We held forums, published reports, testified at public hearings, and shared our conclusions.

These efforts generated studies about various aspects of the tragedy, including what would be the cost of 9/11 to New York's economy, what could be the future mix of economic activities Downtown, what transportation investments should be prioritized, and how could the community be knit back together by good urban design? With support from Richard Kaplan and the J. M. Kaplan Fund, we created a video to explain the transportation network in Lower Manhattan and what could be done to make it better than before. These studies, from a variety of academics, business groups, and volunteers, provided an important basis for the planning that was yet to come. Some decisions, such as restoring Greenwich Street as a through-street, were heavily influenced by these early reports.

Stephen Heintz at the Rockefeller Brothers Fund was especially interested in using new techniques and technology to foster public dialogue and inform the planning process—what we were calling "electronic town hall forums." With his support and [that of] many others, including funding from the Port Authority and Lower Manhattan Development Corporation, the Civic Alliance held Listening to the City at the Javits Center on July 20 and 22, 2002. These events were organized by a group called AmericaSpeaks, which had revolutionized the use of instant polling to energize public forums. Almost 5,000 citizens attended the event on July 20. Representatives from the public agencies and planning firms presented different alternatives for rebuilding the World Trade Center site, and participants were able to provide instant, direct feedback.

We provided daycare, translation, and transit services to get a diverse crowd, and we had specific targets for registrations—how many from the Bronx, how many from New Jersey, how many blue-collar workers, etc.—because we wanted to make sure that the attendees were demographically representative. We had a skilled convener in Carolyn Lukensmeyer and had rented little transponders to allow everyone attending to register their opinions after the different plans had been explained to them.

On the table for discussion were the first proposals from Beyer Blinder Belle. Even though the questions about the proposals were written to favor a positive response, the response on that first vote came back overwhelmingly negative. There was a little bit of a pause, and then, quite literally, a roar went through the Javits Center. It wasn't completely fair to the planners—they were simply doing exercises to explain how the density could be arranged on the site. But it was the first time that the public felt that they had a chance to register their concerns, and they wanted to be heard. Also, the public agencies were insisting that the plans replace all the lost office space, and many people wanted to see the mix of uses at the site change, so this was their opportunity to voice those opinions.

At the Javits Center meetings, when the attendees overwhelmingly refuted the first designs for the WTC that had been put forward by Beyer Blinder Belle for the LMDC, I was standing next to Mr. John Whitehead, the Chair of LMDC, who did not expect that response. About a minute later, he leaned over and said, "I think you've just saved us a couple of years on this process."

The public meeting was initially a setback to the planners at LMDC and the Port Authority, and they quickly reconsidered their strategy for the generation of plans for the site and methods of evaluating designs for rebuilding.

LMDC's New Design Competition

After the Javits Center events, Bob Yaro met with Lou Tomson and Alex Garvin of the LMDC and Joseph Seymour, then Executive Director of the Port Authority. Seymour still wanted to keep the Beyer Blinder Belle design, but Alex Garvin moved quickly into the void created by the public reaction, and he proposed a design competition to open up the planning to more innovative approaches. Alex Garvin remembers:

> In response (to the Javits Center Meeting), the LMDC restarted the planning process, this time requesting participation from designers from around the world. That September, more than 400 design teams from five continents competed for the opportunity to plan the rebuilding. Seven were selected to work with the LMDC and the Port Authority on devising proposals for rebuilding. More important, the competition established the precedent of seeking the finest design talent from around the world, a precedent that would be followed by the Silverstein Organization, the Port Authority, and the LMDC in selecting architects for all rebuilding at the World Trade Center. (Alex Garvin, panelist, T+15 symposium session on "The Memorials, the Public Realm, and Downtown as a 24-Hour City," May 9, 2017)

To provide flexibility in the selection of a winner, rather than being called a competition, the effort was launched as the Innovative Design Study. An Advisory Panel was formed, and the initial phase, a request for qualifications, began with responses due in September 2002.

A critical change in the process of addressing the rebuilding of the site occurred with this new effort. Instead of providing a defined program as had been done previously, only a set of general guidelines was offered in the hope of being less constraining and stimulating more creativity. These included the following:

- Create a tall symbol or structure that would be recognized around the world.
- Avoid building on the footprints of the original towers.
- Restore the original street grid and connect to Battery Park City.
- Centralize transit facilities and provide public open space.
- Flexibly accommodate the provision of commercial and retail space, as determined by the Port Authority.

More than 400 design teams from around the world submitted their proposals. The breadth and variation in response provided a necessary expansion of perspectives on the rebuilding. Reviewing the submissions was a gargantuan task, not only for the LMDC but for the Advisory Panel comprised of professionals in architecture and urban design, representatives of the victims' families, and other community-based stakeholders. Roland Betts, the LMDC Board Director acting as the Chairman of this group, drove an urgent review process, and six teams were selected to work for three more months on the detailed design study. These teams were:

- Daniel Libeskind
- Norman Foster

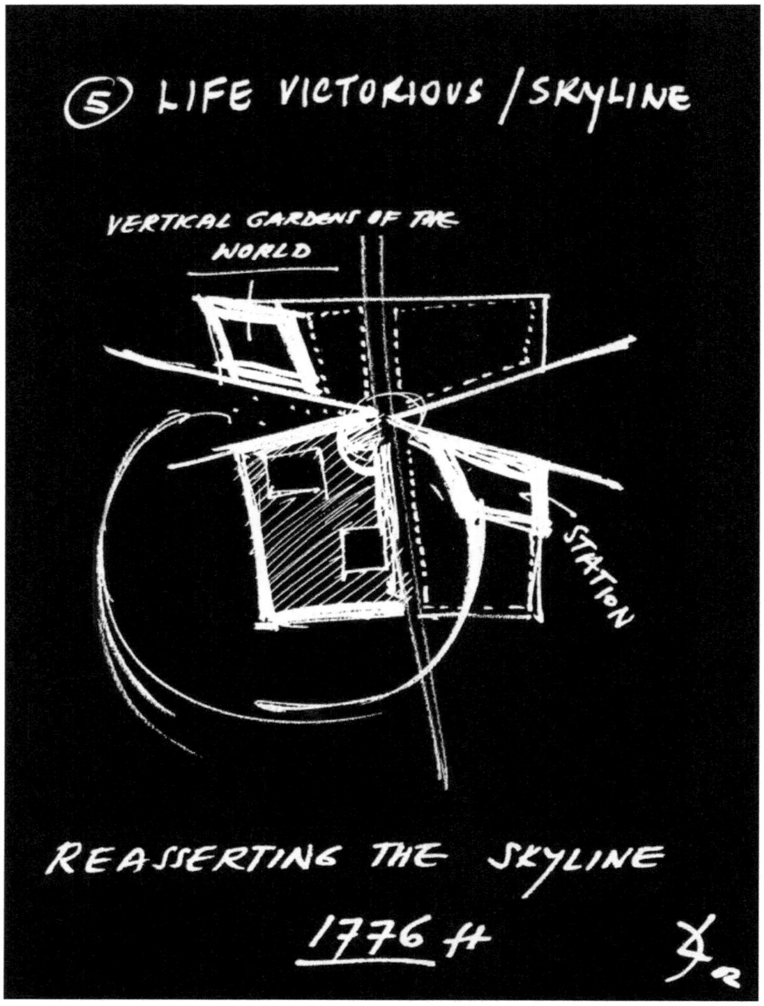

5.3
Concept sketch of the Daniel Libeskind competition submission.
Source: Studio Libeskind/Silverstein Properties.

- Richard Meier, Charles Gwathmey, Peter Eisenman, and Steven Holl
- Rafael Viñoly, representing the team THINK
- Greg Lynn, representing United Architects with Imaginary Forces, a computer graphics firm
- Skidmore, Owings & Merrill (SOM)

Betts, in leading the selection of designers, referenced the direction of his longtime friend, President G. W. Bush: *"Do something that will make people proud!"*

Public Presentation and Commentary

The venue selected for the December 18, 2002, presentation of the work of these six design teams was the translucent public place, the Winter Garden in the World Financial Center (WFC), designed by César Pelli. There had been a celebration of its reopening in the early fall of 2002, and its rebuilding was regarded as a strong symbol of progress.

Paul Goldberger (2004) provides a detailed narration of the event, and he captures the tones (emotional and professional) of the contestants as well as those of the hosting organizations and the audience itself. He concludes:[4]

> But no one really knew at that point what winning the competition would actually mean. It would be several weeks before the LMDC figured out which of the architectural teams would be designated as the official master planner of Ground Zero, and the agency was making no promises to actually build the design of the team that was selected. The Port Authority and Larry Silverstein [owner of the office ground lease] still had to weigh in, and the LMDC itself might come up with plenty of other ideas as the planning process evolved, so in reality, the competition was more of a starting point than a conclusion.[5]

The Critics and Their Suggestions

Given such a high-profile project, many architectural critics commented on the process, the architects, and the plans, and many important ideas were brought into the public dialogue.

The urbanist and academic Michael Sorkin was critical of the lack of public inclusion in the competition procedures and called for "an open process, a genuine competition, in which public bodies (not to mention architectural critics) devoted themselves to promoting the widest—and wildest—styles of inclusion..."[6]

Herbert Muschamp, the Architectural Critic for *The New York Times*, covered other submissions of design decisions, particularly those from a competition run by New York New Visions, in the newspaper's magazine in September 2002. One entry that he highlighted as being a creative and competent solution was by Frederic Schwartz, an architect who had previously worked on solutions for Manhattan's West Side Highway; his design buried West Street as a tunnel and placed all of the required commercial program overage over it, leaving the WTC site completely free for a memorial.[7]

No clear directions were garnered from these alternative architectural proposals. In fact, Paul Goldberger[8] viewed them as exhibiting a

> confused notion of how to give architecture more prominence in the rebuilding process. ... Aesthetic ideas came first: a particular aesthetic vision was paramount, and it was to be imposed on the site before other decisions were made. Whatever society might truly want or need for the site was expected to fit into it.

The Choice of Master Planner

The period in which alternative projects were presented and reviewed allowed for some key aspects to coalesce in the collective minds of the audience and the competing teams—ideas that filtered through the succeeding design development and have come into effect.

On February 4, 2003, based on public reaction and weeks of analysis, the Port Authority and the LMDC decided to continue working through the planning details with the designers on two submissions: "Memory Foundations" by Studio Daniel Libeskind and "World Cultural Center" by the THINK team.

Then, on February 26, Governor Pataki announced that the Studio Libeskind proposal would provide the basis for a master redevelopment plan. This proposal included restoring many of the streets that had been eliminated by the original WTC, the creation of new public places, a major transit station, cultural buildings, and 10 million square feet of commercial space.

Reflections of Daniel Libeskind,[9] Architect of the Master Plan for Rebuilding the World Trade Center

My original idea of New York was as an immigrant. I came here at 14 and saw the Statue of Liberty and the great skyscrapers of Manhattan, and as a student saw the World Trade Center being built. So my first thought about Ground Zero was: Never build on the Tower's footprints—not in the center of the site, which is the location of the memory. Then, whatever we build must reflect the proximity to the Statue of Liberty and reflect the optimistic and vital spirit of New York City. I thought we must put all of the rebuilt buildings at the edges of the site. And then, for the footprints of the Twin Towers, I thought of waterfalls, which would screen the city noises and give space for the sense of memory and contemplation. I also thought of the "Wedge of Light," first as an additional entrance from Church Street, then through the roof on the original Calatrava design for the Oculus. Remember that in the original design, the wings were designed to open the roof to let in the light at 8.46 a.m. and 10.28 a.m. on the day annually to mark the collapse of the Twin Towers.

I had wished to retain the slurry walls, reaching down to the bedrock, to retain the visual sense of the site and as an element of orientation. However, the site has been returned to street level, where it connects to the neighborhood at the same time as providing a grouping or cluster of skyscrapers that are formally connected by height and create their own place. Although I was aware of the economic need to build the skyscrapers, I did not want to create mega structures. Therefore, undulating across the site, I built the towers with the larger footprints on the edge to merge with the other office buildings on Wall Street. But I also created compelling permeability by not rebuilding the "platform," but rather opening up the site to the neighborhood streets. With the addition of other uses on the site, such as retail, the museum, the transport connections, and the Memorial, I was able to create a neighborhood that has drawn over 200,000 people back to Lower Manhattan and the former World Trade Center site has become a new hub of vitality and activity for New York City.

The combination of the public space and commerce, and the balance of memory and optimism, was something that I thought would resonate with New Yorkers, such a unique character of the rebuilding effort. Basically, it is the balancing of the Memorial with a lively commercial hub. Though one must proceed cautiously

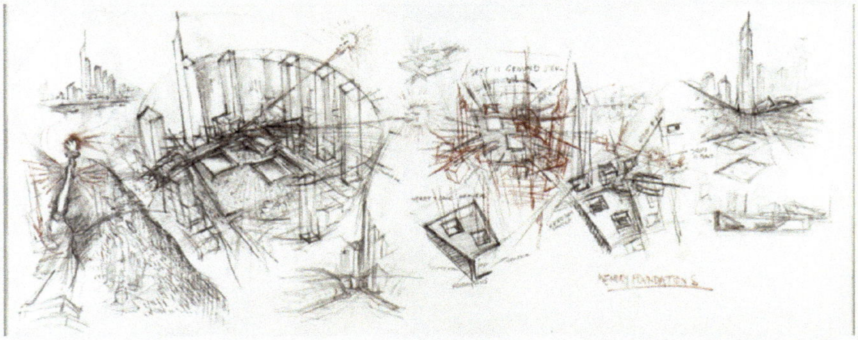

5.4

Libeskind concept for WTC site.

Source: Studio Libeskind.

since it is easy to erase memory, to build bluntly, but we have avoided that, and I am pleased with the balance we achieved.

The general concept of the project is for the site to function as a spatial hinge—a hinge that connects the 10 million square feet of new active space with the Memorial for contemplation and remembrance. Although there are stores on the site, they do not occupy the center but are arranged around the periphery to link with the commercial activity of the city streets of the neighborhood.

To encourage the public to the site, we have created open spaces and thoroughfares that are welcoming and easily navigated—there is a balance of 50 percent open space versus 50 percent buildings. In this way, while respecting the notion of the hallowed ground in the footprints of the original towers and in the reflecting pools, we have inserted the vibrant heart of New York City in the remaining open space.

On the Design Process:

This was not a design competition—it was a Master Plan competition—though it became a [design] competition in the end. However, it was very important that it be a public process. Sure, it is noisy, and you have to listen, although you don't necessarily have to do everything they say: You have to distill it. In listening to the public voices, I felt the need to not only remember the tragic day but also to avoid creating a dark picture that would not be in keeping with New York. Resolving the wishes of many, many stakeholders was also a significant challenge, but I was determined to include a consensus of spirit and feeling in the outcome.

I disagreed totally with those who said, "Rebuild as it was," or with those who said, "No buildings at all," or even Mayor Giuliani who wanted "only low buildings." Manhattan is about density!

In the beginning, the Port Authority had a different idea of the future from those singular visions expressed by the public officials and commentators: They wanted to mix and match, but this was changed during the process.

5.5

Libeskind Master Plan.

Source: Studio Libeskind.

We approached the Master Plan thinking about the underground areas: how to rebuild the subways and PATH and then bringing back the street grid at ground level. We had the footprints to be the Memorial, and we planned where the Performing Arts Center was to be built, and this has all happened.

5.6

Daniel Libeskind, October 2014.

Source: Joe Woolhead.

Notes

1 The community boards of the NYC government are appointed advisory groups for community districts in the five boroughs, and advise the NYC Council on land use and zoning in their respective neighborhoods.

2 Paul Goldberger, *Up from Zero* (Random House, 2004), p. 102.

3 Then President of the Regional Plan Association (RPA).

4 Goldberger, *Up from Zero* (2004), p. 18.

5 And at the closing of the day, Goldberger noted the comment of Lou Tomson of LMDC: "These are among the greatest minds in the world. To reject what they've offered is to say the problem is insolvable."

6 Michael Sorkin, "Making Lists: The Byzantine Politics of Picking Design Firms for Ground Zero," *Architectural Record* 190(11), p. 63.

7 Herbert Muschamp, "Don't Rebuild. Reimagine," *The New York Times Magazine*, September 8, 2002.

8 Goldberger, *Up from Zero* (2004), p. 119.

9 Daniel Libeskind is an international figure in architecture and urban design, renowned for his ability to evoke cultural memory in buildings.

Chapter 6: A Memorial for the Lives Lost and a Museum of the Tragic Event

The immediate and most critical suffering from the devastation was that of the families and relatives of the 2,977 innocent lives lost, and the first responsibility was to memorialize those victims and care for the families and loved ones they left behind.

Support for the Victims' Families

To formally establish a support structure for the families of the victims, the LMDC established a specific space for the families to gather and receive care and comfort. Adjacent to the LMDC offices, this facility became known as the Family Room.

Allison Blais joined this effort in its early planning phases.

Reflections of Allison Blais[1]

Some of my most vivid memories of the days after 9/11 are of the tributes and memorials that seemed to pop up instantaneously, blanketing the city landscape. Any place where people could gather—parks, subway stations, churches, firehouses, even storefronts and restaurants—became a place of remembrance. I remember walking along the Brooklyn Heights Promenade in the days after the attacks, staring at the broken, smoky skyline across the river, locking teary eyes with strangers in a way I never had before. The railing was covered with homemade tributes: a cloudy painting with the words "The rain is the tears" in a child's hand, a billowing bedsheet announcing in scrawled paint that "New York is still standing." The need to memorialize was urgent, and at its core was the need to communicate our shared loss, to connect, to come together.

While that impulse was immediate, as the Reflections throughout this book attest, the journey to realizing the Master Plan for the World Trade Center site and the 9/11 Memorial & Museum at its core was a complicated one. Alongside the public mandate to rebuild, and rebuild quickly, were—of course—deeply passionate opinions on all sides about *what* to build. Unlike many other memorial sites, this one is smack in the middle of one of the busiest cities in the world—and it's on truly sacred ground, where thousands of lives were taken through unthinkable violence.

DOI: 10.1201/9781003476139-6

So there was naturally a vast array of people intimately connected to each decision that needed to be made, large and small: family members of those killed, survivors, first responders and recovery workers, and government officials weighing the needs of our city, state, and country, the business community, residents of the surrounding neighborhood. The challenges facing the Lower Manhattan Development Corporation (LMDC) loomed large, echoed by those thousands of voices with intensely felt and often opposing views, equally valid. One of the first things I learned about my colleagues was the genuine importance they placed on listening to the ideas and opinions of those most closely impacted by 9/11. Making recommendations and decisions can be excruciating when hearts are at stake. As in any job, you can't satisfy everyone, but in a job like this, you wish you could with all your soul.

A few years ago, as we passed the World Trade Center, my taxi driver, without any awareness of my profession, said: "It's easy to destroy, not easy to rebuild." It's true, and there were times when making progress felt impossibly hard—when newspapers bemoaned "the pit at Ground Zero" on a daily basis, or when political and budget battles threatened to devour arguments over what to build and how, and all the while victims' families battled grief of such depths I struggle to imagine.

At those times, I'd retreat for a few minutes to the Family Room—a private space for the relatives of 9/11 victims that my colleagues and I from the LMDC, and later the 9/11 Memorial & Museum, maintained. Its windows overlooked Ground Zero, but you never would've known it since the whole space, floor to ceiling, was plastered with photographs, drawings, dried flowers, and cards for missed birthdays. It was a tangible reminder of the sacred obligation and absolute honor of being part of this project.

The Memorial

Anita Contini was appointed a Vice President of LMDC and quickly established the WTC Memorial, Cultural and Civic Programs initiative to oversee the planning of the Memorial and cultural facilities at Ground Zero.[2]

For the Memorial assignment, Anita first worked with the Families' Advisory Council to create a mission statement, then took members of that Council to view memorials and sites in NYC, Washington, DC, Pennsylvania, and Oklahoma City. By late spring 2002, she launched a competition for the design and vision of the Memorial and selected a jury of 13 prominent professionals, including one family member representing the victims.

Looking Back, Anita Contini Reflects

I accepted the request to join the Lower Manhattan Development Corporation to set up a process that would choose a memorial. I began my efforts in the spring

6.1
Remembrances of the victims.
Source: Joe Woolhead, courtesy of Silverstein Properties.

of 2002 and continued to manage the Memorial and Cultural Program until 2005. An early issue that concerned us was, after such a shattering event, when would be the best time to develop a memorial? The horror and loss were still vividly raw with family members and survivors. Some felt it too soon to start the process and argued that years would be required before there was enough distance from the pain and sadness to make fully considered decisions. Others felt greater urgency, believing a meaningful and respectful memorial was essential to overcome the trauma and heal the scars left by the attack and that any delay would only prolong the pain. In the end, the cry for speed won out, and while site cleanup was still underway, we established criteria and processes for managing decisions about the site and the Memorial.

From its inception, the Memorial process was inclusive. It involved holding community "town halls" and meetings for residents and stakeholders, tours, [gathering] input from other memorials around the country and public comment on emerging design objectives and criteria. After these efforts, we announced a design competition that was open to all and named a 13-member jury to evaluate submissions. By the time the open call for proposals closed, 5,201 design proposals had come in from around the world. This response may have seemed overwhelming at times, but the jury, which had committed to reviewing every proposal, followed a carefully structured and thoughtful review process. It took from August 2003 to early January 2004 to make a final decision.

to help others—are heartbreaking. But to me, they also represent the ultimate reminder that it's the connections in our lives that matter most.

And those connections transcend even what we'd thought about or expected, as we saw on the day the Memorial opened to the families on the 10th anniversary of the attacks. Because of the way the names are arranged, no matter where you looked that day, you saw people coming together. You saw families reuniting, paying tribute to the two cousins they lost whose names were now side by side in perpetuity. Members of a firehouse linking arms as they mourned their lost brothers. Even strangers who'd never met but who knew that their sons' names had been inscribed next to one another because they had been best friends at work took a moment to introduce themselves and embrace. This names' arrangement, based on connections between the victims, also connects those who remain and endure.

Seeing those families coming together on September 11, 2011, knowing the connections between the names, is a memory etched in my heart—along with so many others who continue to this day on this incredible and humbling project. Those kinds of moments, at the very site of the attacks, feel fiercely defiant and full of hope. I think back to the Memorial opening often as I pause at the names, especially at night, when light fills the voids of their absence.

6.4
Names of the victims surrounding the Memorial Pools.
Source: Joe Woolhead, courtesy of Silverstein Properties.

National September 11 Memorial & Museum

Reflections of Alice M. Greenwald[4]

When I arrived in New York in the spring of 2006, the project to build the Memorial and Museum at Ground Zero had been transferred from under the aegis of the Lower Manhattan Development Corporation (LMDC) to the then-called World Trade Center Memorial Foundation. Up until that time, I had superficially followed from afar the evolution at the site, mostly through media reports on the Libeskind Master Site Plan and the Memorial design competition. Perhaps because of my professional orientation, I was more acutely aware of the turmoil surrounding the cultural aspects of the redevelopment process—in particular, a decision to halt the planning of a museum project known as The International Freedom Center.

The tensions underlying that decision—a fundamental incompatibility of planners' goals and content orientation with stakeholder anxieties and aspirations, exacerbated by political imperatives and mostly by how recently the events had taken place—were instructive, as were my nearly two decades of engagement on the Holocaust Museum project, first as a consultant during its formative planning and design phase through to opening; and then, serving as a technical expert during its initial years of operation. I brought to this new project (in New York City) a keen awareness of the power of storytelling in a museum setting to forge emotional connection; the immediacy of artifacts, documentary imagery, and archival materials to convey authenticity; the requirement, given the specificity of this museum's location at a site of atrocity, that we focus intently on what had happened here; and, the imperative of listening to the perspectives of a broad array of constituents and integrating them into our planning process.

I also arrived at a moment of significant change in the project itself. Among the very first conversations following my appointment as Museum Director were several I had with Frank Sciame, a leading construction executive, who had been charged by Governor George Pataki and Mayor Michael Bloomberg with developing a more financially and operationally feasible design for the Memorial that would still maintain the power and integrity of Michael Arad's core design concept that had so moved the competition jurors. Through these conversations, I began to appreciate more fully both the challenges and opportunities of this project. Frank's recommendations, drafted with input from the LMDC and ultimately accepted by its board, eliminated underground galleries and brought the full experience of the Memorial to grade. That decision opened up a cavernous space for the Memorial Museum to be located directly below the Memorial Plaza, enabling the museum experience to include meaningful access to *in situ* archaeological remnants of the original World Trade Center. Frank's plan also transformed what had previously been a separate "visitor orientation and education center" into the Museum's entry pavilion.

It was in this context that I brought on board a museum design team comprised of a curator, creative director, historian and memorialization expert, museum educator, oral historian, and exhibition developer to begin imagining what would ultimately become the 9/11 Memorial & Museum. But we hardly worked in isolation. Among the very first things we did, starting in the summer of 2006, was to convene a "museum planning conversation series" to which a large and diverse group of stakeholders were invited. Our goal was to ensure that the visioning and evolution of this museum would be a collaborative effort and a shared journey, one that demanded close coordination with, among others, the executive leadership of the Memorial Foundation and its Chairman and Board of Directors; the project architects; our colleagues managing the architectural design and construction effort; landmark preservationists; representatives of various constituencies (among them, family members of victims, survivors of the attacks, first responders, former recovery workers, and Lower Manhattan residents and business owners, all with a vested interest in what this museum should and would present); government officials; and the exhibition designers and media producers charged with translating all of the various imperatives into a cogent and meaningful experience.

Our challenges were legion. The work to create the Museum took place within the context of intense public scrutiny, divergent expectations of what would be appropriate to present at such an emotionally charged site, and the daunting responsibility to construct an exhibition narrative that would effectively codify a history not yet written. Particularly challenging among the issues we faced was how to balance the dignity of commemoration with the imperatives of education, historical documentation, and the presentation of information that could be both graphic and provocative. Adding to these complications were our temporal proximity to the event itself; the fact that key constituencies remained traumatized by grief, both personal and communal; and the extremely public, and at times, politicized, planning process for a museum commemorating an event at once highly local, distinctly national, and essentially global.

All told, the project would take eight years to realize. Apart from the inherent challenges of developing a comprehensive storyline, securing artifacts and loans, craning multi-ton artifacts like the steel tridents and last column into an underground site, and securing hard-won consensus about our approach and design strategies, we faced at various times unexpected impediments. These included a full year's moratorium on construction, and then, while the Memorial and Museum were still situated in an open construction site, the flooding of its underground spaces during Superstorm Sandy, which—miraculously—did not result in any permanent damage.

As the Museum began to take physical shape, we were determined to make a virtue out of the necessity of balancing what is essentially an archaeological site with both a powerful memorial obligation and a museum's responsibility to tell a story. We knew that to be successful, we had to make the experience feel seamless and completely coherent. This was hardly a given. With two architectural teams on the Museum project specifically and two for the Memorial (Snøhetta for the aboveground museum pavilion; Davis Brody Bond for the below-ground museum spaces;

Michael Arad of Handel Architects and Peter Walker Partners for the Memorial); three exhibition design firms (lead exhibition designers Thinc Design and Local Projects, a media design and production firm, plus the historical exhibition designer, Layman Design); and two fabrication firms (Design & Production and Hadley Exhibits), not to mention the oversight of the Port Authority of New York and New Jersey, which, in addition to being the owner of the site, assumed construction management responsibilities, collaboration was mandatory, and compromise was inevitable.

But, in telling our story, we never compromised on certain core commitments. We knew that our exhibitions had to provide a sense of historical context. This tragedy was not the result of a natural disaster. 9/11 was caused by human agency, and it was imperative to address who did this and why, even over the objections of some stakeholders.

We did not shy away from difficult subject matter. We sought to establish a level of literacy about the historical context for the events of September 11, the nature of the world in which we live, the reality of terrorism, and the often incompatible political forces that remain in tension with one another.

And we had no choice but to acknowledge that this story is not over yet—that our collective understanding of 9/11 will continue to evolve over time.

We anticipated that a significant percentage of our visitors would be coming here—to this site, to this institution—as a kind of pilgrimage, in an act of bearing witness. Now, more than 20 years since the 9/11 attacks, it is the members of a generation with no living memory of that day who are now coming here to learn about what is, for them, already past history. To reach and be meaningful for both audiences, this museum needed to tell its story in a way that could be felt as much as known, not so much the history of an event as the human experience of that historical moment.

Throughout the process of creating this museum, I remembered my outcry on the day of the attacks: "Oh, the people, the people." This is not a museum of terrorism. It is a storytelling museum about the very real impact of this particular act of terrorism on the lives of very real people and their families, friends, colleagues, and communities. The story it tells is about people; people like us—who got up in the morning and went to work and got caught in the vortex of global events.

By focusing on the human story, the Museum has become a moral platform attesting to the indefensibility of terrorism, to the absolute unacceptability of indiscriminate mass murder as a political response. It is a place for understanding ourselves and the world in which we live. Long before the Museum opened, we would say that our vision was an ambitious one, but the stakes were far too high not to try to achieve it. I would often quote a scholar, Professor Kristin Ann Hass,[5] who wrote that memorials are the way "people make promises to the future about the past." Today, as millions of visitors from all around the world pass through our doors, the 9/11 Memorial & Museum asks each of us to imagine the world we want to bequeath to our children and grandchildren. This museum stands as our collective commitment to the promises we are willing to make.

6.5
President Obama opening the 9/11 Memorial & Museum.
Source: Joe Woolhead, courtesy of Silverstein Properties.

Issues with Construction of the Memorial

In early May 2006, the construction management firm Bovis Lend Lease, who had won the building contract for the Memorial and the Museum, reported that the estimated cost of the two projects had risen to $970 million, not including the multimillion-dollar costs of supporting infrastructure. This was almost twice LMDC's original estimate of $500 million. *The New York Times*, on May 6, 2006, called for a comprehensive review of Bovis's cost estimates and for the Port Authority to take over building the Memorial and Museum.

In 2006, Frank Sciame[6] was appointed by Governor Pataki and Mayor Bloomberg to review the budget and design of the World Trade Center Memorial and given 25 days to report back with a viable plan that aligned more closely to original estimates and therefore ensure a buildable Memorial. Sciame agreed to the details of this assignment, though it included a complex authority to direct changes to the design detailing, which was then well underway by the winning team of Michael Arad and Peter Walker. His diplomatic and successful handling of this process provided an important example of how public interests (in the striking design of the Memorial) can be accomplished while managing a financially feasible rebuilding project.

Reflections of Frank Sciame

The major changes to the earlier design included eliminating the planned ramps that would descend underground, past the waterfalls, as well as the planned underground viewing platforms. Considerable debate was swirling around the location for the names of those lost on 9/11, including the names of first responders. My recommendation was that the names should be at the plaza level. All of the design changes were agreed to by Michael Arad and Peter Walker.

We realized additional savings by eliminating a separate museum entrance and instead utilizing the planned Visitors' Center designed by Snøhetta.

My new estimate of cost for both the Memorial and the Museum was $510 million (although that was soon revised to $530 million with the inclusion of some additions). The Port Authority was to be responsible for most of the underground costs.

Frank Sciame's budget and design recommendations were implemented. Frank remembers that he and his advisors "took great pride when the September 11 Memorial and Plaza and the National September 11 Memorial [&] Museum were successfully built and opened on the 10th and 13th anniversaries, respectively, of the 9/11 terrorist attack."

Notes

1 Intimately involved with The Family Room, Allison Blais worked on the World Trade Center rebuilding for nearly 15 years. She joined the LMDC's early planning efforts and then served as the National September 11 Memorial & Museum's Chief of Staff for eight years.

2 Anita Contini served as Director of WTC Memorial, Cultural and Civic Programs from 2002 to 2005; from 2010 to today she serves as head of the Arts Program at Bloomberg Philanthropies.

3 A native of Israel, Michael Arad was raised there, in the UK, the United States, and Mexico. He is an architect and resident of NYC and was working in the Design Department of the New York City Housing Authority during the Memorial competition.

4 Alice M. Greenwald served as President and CEO (2017–2022) and Director (2006–2016) of the 9/11 Memorial & Museum.

5 Kristin Ann Hass, *Carried to the Wall: American Memory and the Vietnam Veterans Memorial* (University of California Press, 1998), p. 38.

6 Frank Sciame is Chairman and CEO of Sciame Construction LLC, which he founded in 1975.

Chapter 7: Infrastructure Repair and Rebuilding of the PATH Station

As the WTC site was fully destroyed, the removal of the massive volume of still-smoking debris and the replacement of vital infrastructure became the first challenges of all major participants in the rebuilding process.

Reflections of Richard T. Anderson[1]

One of our staff members burst into the room to say, "You won't believe this, but a plane has crashed into the World Trade Center." We went immediately to turn on the TV, thinking at first that it must be a small plane. When the next plane hit the South Tower, and it became obvious that the first plane into the North Tower was a large one, one of our Building Congress members at the meeting, a leading structural engineer, said, "Those buildings will stand—they are so solidly built." But as we watched the collapse of each of the towers, it was clear that more terrible forces were at work, such as the enormously high temperature of the jet fuel in the planes, that overwhelmed the solid steel construction.

The construction engineer said, "I'll volunteer to help," and then I called John Cavanagh, CEO of Morse/Diesel (later AMEC) construction company and the former Chair of NYBC. Cavanagh said that he would call leading contractors and their unionized workers in the industry to go down to the WTC site.

Then I said, "I'm also going to call the leaders in the construction industry." Then I received a call from Michael Burton, Deputy Commissioner of the NYC Department of Design and Construction (DD&C), under Ken Holden, asking, "What can we expect from the industry?" and I said, "Complete cooperation," and I was calling the industry leaders.

Everyone volunteered. The site, that smoldering pile, was divided up into four quadrants, each led by four different companies—Morse/Diesel, Turner, Tully Construction, and Bovis/Leher McGovern ... And everything was done on a handshake. The construction companies advanced all of the money needed. Months went by, and no contracts were signed, as the effort went from first rescue, then recovery, and then into the years of rebuilding.

DOI: 10.1201/9781003476139-7

7.1
Smoldering ruins of the collapsed WTC.
Source: Joe Woolhead.

My role was as a voluntary coordinator, working with DD&C. But everyone was a volunteer—the companies, the labor guys, the engineers, the architects.

The industry was magnificent. Everyone contributed. All volunteered. And no one waited for a contract.

Two weeks later, we held an NYBC event for our membership, asking the people who were leading the rescue to tell the rest of the industry what was happening at the 16-acre site. Those leaders told their member colleagues just how tough it was working on "The Pile." Andrew Cuomo, then the NYS Attorney General, was the speaker that day, invited to make a few comments and observations to the members in attendance. Cuomo stressed to them the full backing and appreciation of the New York State government for these amazing efforts.

On 9/11, Ray McGuire, then Managing Director of the Contractors' Association of Greater New York (CAGNY), comprising NYC's major high-rise contractors, was having breakfast with Tim Thornton of Thornton Tomasetti. They heard waiters whispering about a "plane going into the World Trade Center," but only when they had reached Ray's office did they realize that planes had hit the Twin Towers. They watched as each of the two buildings collapsed.

Reflections of Ray McGuire

Tim said to me, "Just a year ago, my team and I went through a what-if exercise, concluding that if a plane ever crashed into the Twin Towers, they would stand; they were so solidly built. But we didn't factor in jet fuel."

I then called John Cavanagh, CEO of Morse/Diesel (later known as AMEC), one of his members and the recent Chair of the New York Building Congress. By that time, Michael Burton, Deputy Commissioner of the NYC Department of Design and Construction (DD&C), had already spoken to Tully Construction, which was completing a project on the east side of the WTC, and to Gene McGovern of Lehrer-McGovern, which had completed several emergency projects for DD&C, and asked them to mobilize at the site. James Abadie, the Senior Field Superintendent for Lehrer-McGovern, was working on a hotel project south of the WTC and was part of the Tully/Morse/Diesel/ Lehrer-McGovern team (later joined by Turner Construction) that convened at 1 Police Plaza on the evening of 9/11.

It was quickly agreed that Tully would assist the first responders conducting rescue operations at the southeast quadrant of what was now being called Ground Zero; that Lehrer-McGovern would control the southwest quadrant; Morse/Diesel the northwest quadrant; and Turner (later joined by Plaza Construction) would work at the northeast quadrant. I considered Jim Amity of Lehrer-McGovern to be one of the many heroes of that daunting effort, "down there 24 hours a day."

I was also going down every day to meet with the firms, as was Ed Malloy, head of the Building Trades Council, who was trying to settle jurisdictional disputes with the labor representatives who had volunteered.

Michael Burton continued to supervise the construction companies involved in the rescue/cleanup for the next nine months. He would hold a meeting every morning at 9 a.m. and every afternoon at 5 p.m. to review what was being done and what next needed to be done.

A few days after 9/11, three environmental specialists from the operating engineers' main office in Washington, DC, came to the site, ran some tests, and declared, "This air/site is toxic; everyone must wear a mask," although I remember that only the members of the National Guard on-site would comply and wear a mask.

One of the many challenges everyone was facing at the site was the arrival of a 100-ton crane, enormous and heavy. Everyone was anxious that this monster might collapse the slurry wall and the Hudson River would pour in and flood everything, including Wall Street. The engineers figured out a way to maneuver the crane to avoid another disaster.

Another challenge was to secure indemnity from NYC to protect everyone working on the site. The city's initial offer of complete indemnity was quickly withdrawn, leaving workers and firms exposed, which led Ray and other industry leaders to ask the federal government for special insurance to protect all the workers. This federal insurance and

7.2
Careful positioning of the crane.
Source: Joe Woolhead.

protection became vital when many thousands of lawsuits were filed by workers against the firms engaged at the 16-acre site.

Despite all those challenges, and many others facing construction and public-sector leaders, Ray McGuire looks back on the remarkable development that now exists on that 16-acre site with great pride in the members of his professional association who worked to make the massive cleanup possible so that development could eventually occur.

The Massive Challenge of Rebuilding the Infrastructure at the World Trade Center Site

The underground area of the WTC is vast—there are 3 million square feet of below-grade infrastructure space. The West Bathtub is 2.4 million square feet, and the South Bathtub is 250,000 square feet. There are 2,500 linear feet of interior roadways and 450+ spaces for vehicle parking.[2]

In her book, Lynne Sagalyn noted that the Port Authority had decided to "rebuild the Trade Center site with an integrated, centralized system underground servicing the entire complex."[3]

Sagalyn and Carla Bonacci list the major elements involved in the underground infrastructure: the Hub, which included the permanent PATH Station, later to become the Oculus, totaled 800,000 square feet—the finished Oculus itself was 42,400 square feet (larger than Grand Central by 28 percent); the temporary PATH Station, which opened in November 2003; and the permanent PATH Station, which opened in 2010. Security concerns added $591 million to its construction cost.

Also listed by Bonacci[4] for the vast infrastructure were the underground work for the Memorial Pools and the 9/11 Memorial & Museum; the #1 Subway line (which had to be dug out, and underpinned to shore up the "box" since Governor Pataki insisted

7.3
Extensive foundations for complex infrastructure.
Source: Joe Woolhead.

the trains be kept running); the final cost from the Federal Transit Administration (FTA) was $355.5 million; the R subway station at Cortlandt Street; repairs to Route 9A; underground construction of 1 WTC; the east-west and north-south corridors linking Brookfield Place to Fulton Transit Center; the East Bathtub excavation, "a jungle of interference from the slurry wall tiebacks"; utilities for the site running under Greenwich Street; placing the central chiller and other mechanical systems (power, telecoms, heating, plumbing, etc.); siting the bus inspection and garage, plus security; siting the PATH vent shaft; and building the infrastructure/base for the Perelman Performing Arts Center.

The Port Authority was given $2.2 billion from the FTA, plus $230 million for extras.

Lynne Sagalyn[5] referred to this challenge as "The Octopus Beneath the Dove." The keyword used by Sagalyn, as well as by Steve Plate at the T+15 symposium session on Infrastructure,[6] was "interdependence"—of the PATH terminal and 20 other WTC projects. In his reflections, Chris Ward, Executive Director of the Port Authority (2008-2011) described the challenge of this process as "a child's game of pick-up sticks—everything was dependent on everything else."

John Gilbert noted:

> Rebuilding and integrating the utilities was a major challenge. Updating of the Lower Manhattan telecommunications systems was overdue at the time, but the absolute destruction required new thinking about what systems were required for the future, i.e., including increased redundancy. (John Gilbert, COO of the Rudin Organization, panelist, T+15 symposium session on "Infrastructure Challenges in Rebuilding the World Trade Center and Downtown Manhattan," May 11, 2018)

In addition, the site was listed as a National Historic Site, accomplished in February 2004 by LMDC. This caused considerable delay as the Port Authority had to find/set aside/get agreement on and placement for artifacts from the site.

Ray Sandiford[7] refers to these new hurdles in his reflections on rebuilding the slurry walls.

Reflections of Ray Sandiford on Rebuilding the Slurry Walls

The WTC subgrade (the Bathtub) has been a topic of marvel since it was built in the late 1960s. Its design and construction pre-dated my Port Authority career, but it played a central role during my 30-year tenure. In the immediate aftermath of 9/11, the stability of its 80-foot-deep walls was a huge concern and led to some drastic action. Their failure could have led to the Hudson River flooding not only the WTC site, but a number of the connecting subway tunnels. Thus, it was imperative that the walls be stabilized as soon as practical. This involved the installation of new tiebacks, consisting of bundles of high-strength steel cables that connect the basement wall to bedrock and thus provide lateral support to counteract soil and water pressures.

WTC Tiebacks Arrangement

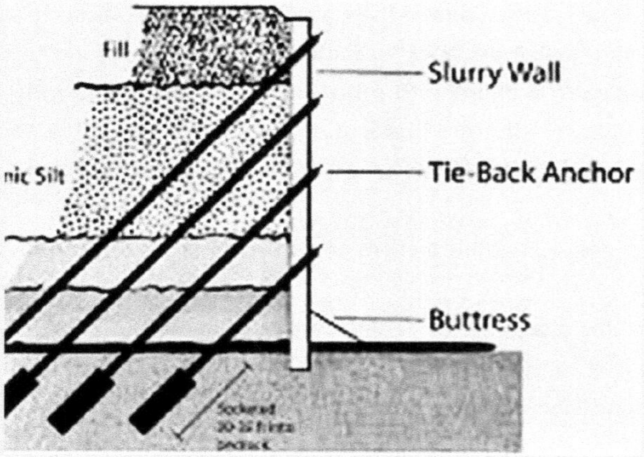

7.4
Tieback arrangement diagram.
Source: Courtesy of Ray Sandiford.

The tiebacks were drilled through the walls simultaneous to the removal of the 9/11 debris and other recovery efforts. This meant drill rigs and their heroic crews perched on piles of barely stable and smoldering building debris. A truly nerve-wracking experience for all involved.

7.5

Tieback construction at the WTC site.

Source: Joe Woolhead, courtesy of Silverstein Properties.

Note on Slurry Wall Construction

The construction of the original WTC basement was a novelty and involved the use of a new technology known as "slurry wall" construction. This technique was developed in Italy in the early 1960s and, in the simplest sense, allows for the construction of permanent basement walls prior to a basement's excavation. Walls are constructed in short, narrow panels, approximately 12 feet long and 3 feet wide and excavated to bedrock. As panels are excavated, they are filled with a clay (bentonite) and water slurry. This slurry is heavy enough to maintain the panels' stability and allow for the placement of steel reinforcing and subsequent filling with concrete. Adjoining panels are connected using end stops to create a near-watertight structure.

The adoption of the slurry wall construction can be attributed to two legends in the Port Authority's history: John Kyle (former Chief Engineer) and Marty Kapp (former Chief Geotechnical Engineer). John Kyle was tapped by Governor Nelson Rockefeller to build the WTC. Governor Rockefeller recognized John Kyle's abilities from his early private-sector days as the contractor who built the complex basement and foundation works at Rockefeller Center. Nelson was the resident engineer for his father (John D. Rockefeller) on that project and thus worked with Kyle on a regular basis. Marty Kapp was an engineer who thrived on innovation, and when the slurry wall concept was presented to him, he recognized its value, and he ran with it. The success of the WTC's basement started an entire new industry and made once-impossible projects possible.

The other immediate action was to seal the PATH tunnels under Jersey City. This was necessary to first prevent the millions of gallons of water the NYC Fire Department was pumping on the WTC rubble from flooding the New Jersey portion of the PATH system, but also to prevent even greater flooding should the basement walls fail and the Hudson [River] spill into the WTC site. The sealing was accomplished with large concrete plugs positioned just east of Exchange Place Station.

The WTC walls were secured in a matter of a few months, and the process of redevelopment planning began. In the interim, though, it was essential to keep the walls stable. Unfortunately, the violence of the towers' collapse caused significant shock, and the concrete of the interior of the slurry walls peeled off, exposing the steel reinforcing. In a matter of months, this steel began to corrode and cause additional concrete spalling. To halt this deterioration and reinstate the required structural capacity, thus, repairs had to be made. This involved water-blasting the remaining loose concrete from the wall and then applying a layer of pneumatically-applied concrete (shotcrete) to the wall. As a side note, although this work was essential for the wall's stability, I was nonetheless berated by the historic preservationist for "destroying an artifact." *C'est la vie*.

With the basement walls stabilized, the site planning would—and understandably—continue for years. When the decision to expand the site into the area east of Greenwich Street was made, plans for new slurry walls were undertaken. There were a number of challenges related to that expansion and the most challenging involved NYC subway tunnels. The single most difficult challenge was constructing under the No. 1 Line running down the center of the site, under Greenwich Street. This involved the underpinning of approximately 1,000 feet of subway tunnel that would eventually be elevated 70 to 80 feet in the air. Complicating the construction was the need to connect the old and new basements under the #1 Subway tunnel at both the north and south sides. This required top-down construction with a complex sequence of ground stabilization (jet grouting), partial excavation, partial wall construction, and then tieback installation. This sequence was carried out until the excavation reached bedrock.[8]

The other significant challenge was basement wall construction along Church Street. Under Church Street, the NYC N&W line station and tunnels are immediately east of the basement. Also, a remnant wall of the original Hudson & Manhattan (H&M) train station is located on the exact alignment of the new east basement wall.

7.6

Construction for the subway tunnel.

Source: Joe Woolhead, courtesy of Silverstein Properties.

The remnant H&M wall appeared to be a real obstacle to construction until investigations revealed it was in better shape than when it was built—note: Concrete gets stronger in time. With that knowledge, it was decided to recycle the wall and make it a part of the final construction.

The station and tunnels under Church Street were another matter. Although the new construction did not impinge on these structures, tieback anchors had to be installed beneath them to secure the new WTC basement walls. As luck would have it, the soils under Church Street are notorious in NYC and have their own nickname—"Bull's Liver." This soil was created during the last glacial period and consists of fine silts and rock flour, which, when saturated, turn to fluid with any vibration. The Bull's Liver lived up to its reputation at the WTC. During tieback installation, these soils "fluidized" and literally shot out the drill casings and caused the station and tunnels to settle. The NYC Transit Authority (NYCTA) was generous in initially allowing us to settle their structures a half-inch, but after settlement reached two and a half inches, things became a bit heated. Eventually, it was decided to get the basement walls built and come back and reinstate the missing soil. This was done with a program of compaction grouting, injecting concrete through the station, and recompacting the loosened soils. This was successful but costly at approximately $10 million.

The new basement expansion was very time-sensitive due to a lease turnover agreement and penalties for late turnover. This resulted in the construction being a 24/7 operation. This accelerated the work but at a cost. The basement expansion cost approximately $250 million in direct expenses, but probably more like

$500 million if overall project delays are considered. In total, the cost of rebuilding the WTC site came to approximately $17 billion, and that included creation of some 15 million square feet of new space. Of that, the substructure facilities totaled some 3 million square feet, providing space for vehicular parking and screening facilities, the new PATH Station, WTC Memorial spaces, a new chiller plant, and numerous pedestrian passages.

There were numerous players in the WTC recovery and reconstruction. I would like to mention a few: MRCE (Mueser Rutledge Consulting Engineers) was a key contributor to the recovery effort, providing expedited design solutions for the emergency tieback installation and slurry wall repairs; the Downtown Design Partnership (an STV/AECOM Joint Venture) led the design of the #1 Subway tunnel underpinning and the Calatrava Station, while the PANYNJ Design Division led the design of the WTC basement expansion and had the privilege to oversee all the other efforts.

Looking back, all the sleepless nights were worth it. No matter how many times I enter the new WTC, my heart skips a beat. To have the WTC rise from those ashes is a near miracle, and I take a little pride in having played a small part in it.

Rebuilding the PATH Station (Twice)

One of the immediate—and major—infrastructure challenges was to repair the PATH site and build a temporary station for PATH commuters to the WTC station.

Carla Bonacci, in her position as Deputy Program Director of Major Capital Projects for the Port Authority from 2002 to 2006, managed the immediate site recovery projects, including the WTC site transition and restoration. She supervised the design and initial construction work for the temporary WTC PATH Station, a $300 million project that opened to the commuting public in November 2003. Then, she oversaw the planning and conceptual development of the permanent PATH Terminal (aka the WTC Transit Hub), including the federal government's environmental review as required by the National Environmental Policy Act (NEPA). In her capacity as Assistant Director of World Trade Center construction, in the spring/summer of 2002, Carla Bonacci collaborated with other senior-level internal, public-, and private-sector executives to transition the WTC site back to the control of the Port Authority; she later organized and managed project teams to plan, design, and build the new underground chiller plant for the complex.

After the removal of the debris, the site represented a blank slate for new planning. Presented with this opportunity to replan the urban fabric of the 16-acre WTC site where Greenwich Street had been removed as a thoroughfare in the original design, Alex Garvin and many of his planning colleagues at the LMDC felt that Greenwich Street should again continue through the site. David Dunlap reported an important affirmation of this intent by Amanda Burden, then the Chair of the New York City Planning Commission: "Greenwich Street is not only symbolically important, but it is an important first step. It's about beginning to interconnect the totally disconnected elements of Lower Manhattan."[9]

This planning was interconnected with the design for the permanent PATH train station, the construction of which required substantial infrastructure work below ground while also impinging on the work being done for the Memorial Pools and foundation of the 9/11 Museum.

Carla Bonacci participated in the request for proposals (RFP) process that selected Santiago Calatrava to design the Hub—which, of course, became the Oculus, when that great space opened in 2012.

Alex Garvin summarized the complexity of rebuilding the transportation infrastructure and the necessary interconnectivity between the various modes:

> When it opened in December 2009, the new station featured an underground passageway connecting the PATH terminal to the ferries at the World Financial Center and more than a dozen subway lines under the WTC site and the Fulton Street Transit Center (which itself is due to be completed in the summer of 2009). The two Cortlandt Street subway stations at the WTC site were badly damaged in the attack and closed temporarily. On September 15, 2002, service returned to the western station; the #1 Subway now passes through as it goes to or from its terminal at South Ferry, but the station will not serve as a stop until the WTC reconstruction is complete. On the same date, the eastern station reopened, serving the R, W, and N trains, only to close again in August 2005 to facilitate the construction of the underground connection between the new transit hubs. (Alex Garvin, contributor, T+15 symposium session on "Infrastructure Issues in Rebuilding the World Trade Center and Downtown Manhattan," May 11, 2018)

Traffic and Roadways

At the time of the devastation on September 11, 2001, Tim Gilchrist was the Strategic Planning Director of the New York State Department of Transportation (NYSDOT), based in Albany. His responsibility was to include the restoration of the state highway Route 9A, which runs between Battery Place and Chambers Street on Manhattan's West Side. The section where it ran along the west side of the WTC, from Vesey Street to the entrance of the Brooklyn Battery Tunnel, was heavily damaged.

Over the following years, Gilchrist directed the team rebuilding Route 9A. In 2007, he joined the staff of Governor Spitzer, and for Spitzer, and subsequently Governor Paterson, he served as a key negotiator on many of the Lower Manhattan issues. Those issues of significant impact on the neighborhood included the siting of the proposed new Goldman Sachs headquarters at Battery Park City, the relocation of the Freedom Tower (the rebuilt 1 WTC), the dismantling of the Deutsche Bank Building to the south of the WTC site, and the critical 2010 Agreement between the Port Authority and Silverstein Properties to resume construction at Ground Zero.

Tim Gilchrist Reflects

Following 9/11, once the temporary roadway was under construction, NYSDOT created a special group dedicated to the rebuilding of West Street. Located in Battery Park City and comprised of staff throughout the State, the group reported

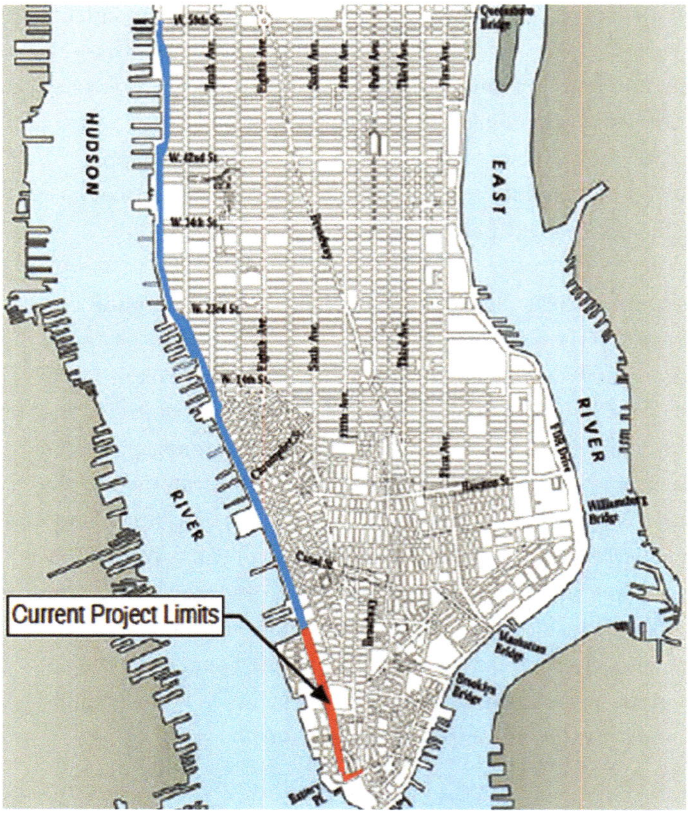

7.7

New York State Route 9A.

Source: New York State government.

to me in Albany. Our work was heavily supported by several consultants' firms, with Vollmer (now part of Stantec) as the primary design consultant.

We needed to address three major issues regarding West Street:

- Should the portion immediately adjacent to the World Trade Center site be at grade or be depressed in a tunnel?
- Could the rebuilding of the section south of West Thames proceed ahead of the northerly section, which would help reconnect Battery Park City?
- Could temporary pedestrian bridges be built to solve the difficult pedestrian access to Battery Park City, pending final reconstruction of the WTC site?

All of the West Street issues were closely tied with other issues at the site. West Street was also the major utility corridor for Downtown. West Street's grade and configuration needed to be consistent with the rebuilding of the WTC site. Our NYSDOT team quickly became part of the multi-agency groups that were formed for transportation and site development and worked closely with the Port Authority, the Metropolitan Transportation Authority (MTA), NYC, the newly formed Lower Manhattan Development Corporation (LMDC), and later the Lower Manhattan Construction Command Center.

Was Route 9A to be at grade or a tunnel?

Immediately after the attack, John Zuccotti, Chairman of Brookfield Properties, led a group proposing that West Street be rebuilt as a tunnel from just north of the Battery Park tunnel to surface between Barclay Street and Murray Street, the eventual site of the Goldman Sachs building. This proposal was fraught with controversy from the start. The Port Authority was worried about how the underground tunnel would affect access to the planned bus depot and truck service entrance on the WTC site. The NYC and other agencies were concerned with the cost and impact on other projects. Battery Park City residents were concerned with the length of construction and access during construction. Governor Pataki directed NYSDOT to begin the environmental process to determine if a tunnel was feasible, and working with the other involved agencies, the Federal Highway Administration and the Federal Transit Administration (administrator of the federal transportation aid for the recovery), we began the process, coordinating with other ongoing environmental analyses, including the future design of the WTC site.

A major challenge during the discussions of the tunnel versus at-grade highway was the location of numerous utilities under West Street. This required negotiations with Verizon, ConEd, Empire City Subway, the NYC Department of Environmental Protection (NYCDEP), and the Port Authority. The Verizon Building at the corner of West and Vesey streets contained the major telecommunications equipment for the financial district, and these lines ran under West Street. In addition to the electrical and steam lines, the NYCDEP had water lines and a 72-inch interceptor sewer line. The water cooling lines for 1 WTC ran under 9A.

During the environmental process, Empire State Development was negotiating with Goldman Sachs to locate their new headquarters in Battery Park City along West Street between Vesey and Murray Streets. Shortly after Goldman Sachs agreed to locate in BPC, they found out that the tunnel was to be the preferred alternative for West Street. Goldman Sachs refused to have the tunnel surface in front of their planned building, saying that would break the terms of their agreement to build at BPC, and gave an April 1, 2005, deadline for a decision on "tunnel or no tunnel." The tunnel issue was not resolved by then, so Goldman announced they would not build their headquarters at this site. Finally, Governor Pataki, speaking at an Association for a Better New York (ABNY) breakfast later that month, announced, "No tunnel," so the Goldman Sachs plan to build went forward.

Once the decision was made not to build 9A as a tunnel, the at-grade configuration required a steady rise in elevation. Each change in the 9A elevation or alignment required changes to the utilities and coordination with all parties. Plans for the site were not yet settled, and an important issue was the eventual elevation of the WTC Memorial and 1 WTC. Their elevation drove that of West Street and the underground utilities.

The new 9A highway, as a result, now slopes upward by 7 feet from the mouth of the Battery Park tunnel to the Memorial, then slopes downward to Vesey Street. This is a gentle grade, almost imperceptible on the street, except for the impact on the walkway in front of Brookfield Place, where the grade difference is noticeable at the curb cuts.

Planning and Building Promenade South

While the environmental process was considering the at-grade and tunnel alternatives for the section of the road immediately adjacent to the World Trade Center site, our team proceeded with the design for the lower section of the roadway, from West Thames south to Battery Place.

One of the first places of remediation was by the exit of the Brooklyn Battery Tunnel, which was perilously close to pedestrians crossing the street. We built a temporary pedestrian bridge at Thames Street to provide a safe crossing. In addition, at the intersection of West Street and Battery Place, southbound cars and trucks had a difficult turn to go back uptown, so NYSDOT's solution was to cover part of the mouth of the tunnel, expand the turning circle and the pedestrian walkway, and install new traffic and walking signals.

We then created the Promenade South, two landscaped passageways, one for pedestrians only, the other a parallel passageway with lanes for both pedestrians and bicycles. This promenade is fully landscaped, with trees and plantings, a community garden, a protected children's playground, and a grassy area for older children and teenagers to play basketball or soccer. The promenade continues up Route 9A to just beyond the new Goldman Sachs Building at Vesey Street. The funds for all of this came from the federal government.

Prior to 9/11, there were two major covered and spacious bridges/walkways over the 9A highway, connecting the WTC to BPC and the ferries at Vesey Street and to the Merrill Lynch/Dow Jones Buildings at the World Financial Center at Liberty Street. The northern bridge was destroyed on 9/11. There were many intense debates about rebuilding the bridges. State DOT's traffic studies showed that fewer people would be using these bridges post 9/11 since many of the large firms had by now moved out, or sharply reduced staff, from the WFC buildings. Also, I worried that because the elevators and escalators from the street would be external, they would never work in adverse weather.

But pressure continued to build temporary bridges for passenger safety, particularly from commuters exiting the ferries and reaching Vesey Street. So, NYSDOT built an additional temporary bridge at Vesey Street at Brookfield's urging. Governor Pataki announced that the bridge would open the same day as PATH service was restored. DOT and the contractor worked furiously to get the bridge constructed in time. Goldman Sachs objected to this bridge and said it had to come down once the Brookfield-Oculus underpass was opened. Now, a new permanent bridge is being built at the southern part, just north of West Thames Street, as a replacement for that temporary bridge.

In addition to the highway-related issues because of the proximity of West Street to the WTC and Brookfield Place, the frontage and bike path in front of Brookfield Place needed to be designed in conjunction with the redevelopment of the WTC as Brookfield Place.

Building for Resiliency

As the devastation in 2001 followed the earlier 1993 bombing of the WTC, the vulnerability of skyscrapers in dense urban environments was given considerable attention. The earlier damage had led to more surveillance and monitoring of such office complexes, but the absolute demolition of the Twin Towers and the destruction of key infrastructure required a rethinking of resiliency and security for the urban fabric across the nation and even globally.

The U.S. Congress commissioned a report to inform national policy discussions regarding critical infrastructure protection and legislated for improvements through the Intelligence Reform and Terrorism Prevention Act of 2004.[10] Many of the recommendations built on the critical infrastructure protection actions that pre-dated the 2001 attacks and were reinforced to counter such terrorist attacks. Procedures for entry into the United States were tightened, and a newly formed agency, the Department of Homeland Security, led intelligence reform and increased restrictions on visas and security screening, particularly for commercial aviation. It reiterated the need for federal policy to:

- identify those transportation assets that need to be protected;
- set risk-based priorities for defending them;
- select the most practical and cost-effective ways to do so;
- develop a plan and budget;
- ensure funding for implementation.

These recommendations also accelerated the implementation of existing policy specifically relating to the organizational structures of government agencies, the integration

7.8
Below-ground-level rebuilding of the new PATH train station.
Source: Courtesy of the PANYNJ.

between federal, state, and local authorities, and the distribution of funding to those entities.

Most importantly, the report and the activities undertaken at the WTC during the rebuilding emphasized the integration of threat information with vulnerability assessments. Further, the planning for and allocation of resources to the protection of critical infrastructure was to be informed by rigorous and ongoing risk assessments that would be functionally and geographically comprehensive rather than merely applied to specific areas, thereby ensuring the formulation of optimal overall risk management strategies.[11]

Notes

1 Richard T. Anderson ("Dick" to everyone in NYC) was President of the New York Building Congress (NYBC), a professional association of construction firms, architects, labor, and related groups, from 1994 to 2017. On the morning of 9/11, Dick was meeting in his conference room with the nominating committee of NYBC and recalls the events of that horrific morning.

2 Detail from a working document prepared by Carla Bonacci.

3 Lynne B. Sagalyn, *Power at Ground Zero: Politics, Money and the Remaking of Lower Manhattan* (Oxford University Press, 2016), p. 482.

4 Detail from a working document prepared by Carla Bonacci.

5 Sagalyn, *Power at Ground Zero* (2016), p. 481.

6 Steve Plate, panelist, T+15 symposium session on "Infrastructure Issues in Rebuilding the World Trade Center and Downtown Manhattan," May 11, 2018.

7 Ray Sandiford was the Chief Geotechnical Engineer at the Port Authority of New York & New Jersey during the WTC recovery and redevelopment. For 30 years, he was involved in most major design and construction works related to the PANYNJ's capital program, and after 9/11, he was involved in and led most activities related to subgrade construction at the WTC and related PATH construction.

8 See 'Daniel Creighton Reflects' on rebuilding the #1 Subway tunnel and station.

9 David Dunlap, "Building Blocks: The Resurrection of Greenwich Street," *The New York Times*, September 26, 2016.

10 U.S. Federal Government Legislation: S.2845—108th Congress (2003–2004)—Intelligence Reform and Terrorism Prevention Act of 2004. www.congress.gov/bill/108th-congress/senate-bill/2845

11 Critical infrastructure protections: John Moteff, *The 9/11 Commission Report and Congressional Response* (Resources: Science and Industry Division, Congressional Research Service, The Library of Congress, January 11, 2005). Moteff is a specialist in science and technology policy.

Chapter 8: Rebuilding Commerce

While the commemorative and symbolic rebuilding of the WTC site and Lower Manhattan proceeded, the challenges continued for reviving economic activity, particularly that housed in the remaining office and retail buildings and those planned to be rebuilt. Although the Lower Manhattan business district continued to benefit from its competitive office rental rates relative to Midtown, the planning battles and slow progress of rebuilding were discouraging businesses from considering it ideal.

Office Space

The terrorist attacks of September 11, 2001, destroyed 13 million square feet of Class A office space at the WTC complex and damaged an additional 15 million square feet of nearby office space, much of which was also Class A quality, mostly located at the World Financial Center (WFC).

By December 2001, 1 WFC, 4 WFC, and 1 Liberty Plaza had reopened, but other buildings required substantial renovation, and the office tower south of the site, 1 Bankers Trust Plaza (also known as the Deutsche Bank Building), would eventually require full demolition.[1]

Amidst the chaos and incomprehension that prevailed in the immediate aftermath came the practical question: Where would the firms that were displaced from the destroyed and damaged office buildings of Lower Manhattan relocate? Fears persisted that all too many of these firms would leave Manhattan and NYC City immediately, and possibly on a permanent basis. However, studies done by the research team at Grubb & Ellis and by Hugh Kelly, then of the Shack Institute of Real Estate at New York University,[2] noted that as early as December 2001, the overwhelming choice had been relocation to Midtown Manhattan or to Midtown South. More space was available in these locations than previously thought as the recession of 2000–2001 resulted in the downsizing of many firms, thus increasing the inventory of Class A space, available either under new leases or as sublet space. Additionally, in this emergency context, the relocating firms were occupying much less space than in their previous Downtown locations.

Kelly has observed that the primary determinants of these office relocations were not just the availability of Class A space, but also the centrality of the large subway and commuter transportation network in Midtown as well as in urban New Jersey. In

DOI: 10.1201/9781003476139-8

Manhattan, the existing presence of interconnected business and professional services in the Midtown area was also a major determining factor in this relocation.

In a post-9/11 analysis prepared for the Civic Alliance to Rebuild Downtown New York,[3] Kelly's research revealed that relocating firms did not indicate price sensitivity with respect to the costs of office space, as the majority of leases taken up were in the higher rent districts of Manhattan's Midtown Core, Midtown West, and Midtown South. A lesser number of firms went to lower-rent areas in suburban New Jersey, Long Island, or the northern suburbs of NYC.

Hugh F. Kelly's Recollections and Reflections on the NYC Office Market After 9/11[4]

The shock of the destruction of the World Trade Center on September 11, 2001, and the reverberations—both economic and social—that persisted for months (and even into the present) color public comprehension on many levels. This is understandably so. For many, this was a "this changes everything" event. Any air traveler can vouch for that. Initiatives such as the "If you see something, say something" continue to press public awareness of potential threats to safety. The nearly universal requirement of identity checks to enter New York's office buildings expands a practice that was only common for a few public buildings, including the World Trade Center, prior to 9/11. Physical barriers such as bollards and other forms of access control can be seen around many office buildings. Less obvious are alterations in HVAC (heating, ventilation, and air conditioning) and other building systems meant to secure against more subtle intrusions directed against building occupants.

However, I am struck by the level of continuity sustained by New York and by the fundamental trends of the Manhattan real estate market since 9/11. Many have remarked about the resilience of the City's economy and its social fabric. That resilience can be viewed as the strength of New York's continuity, not as a return to pre-9/11 conditions but as organic adaptability to changing circumstances—New York as a vital place that is constantly reinventing itself.

This process comes about partly because our economy and real estate market are complex systems, deeply interwoven networks of relationships in civic, business, political, social, and charitable dimensions, networks that provide the capacity to maintain alternative paths of connection when one set of links is disrupted. But the process is not automatic or deterministic. Human systems are the creation of human beings, and our ability to make choices is a powerful factor in change. I believe that crises are times when choice comes particularly to the fore.

Just three months before 9/11, the so-called "Group of 35" report by a group of business, labor, and academic panelists (convened by Senator Schumer and former Secretary of the Treasury Robert Rubin) had laid out a case for continued growth in New York's office stock. The report included anticipated development throughout Manhattan, including the Far West Side and in office submarkets such as Long Island City and downtown Brooklyn. In particular, the report highlighted the aging of Manhattan's existing office stock, a theme consistently sounded by industry leaders such as CBRE's Mary Ann Tighe. The Group of 35 report provided important statistical background for planning in the post-9/11 environment, as well as a vision for an expanding citywide economy.

In the immediate aftermath of 9/11, the Real Estate Board of New York, under the leadership of REBNY President Steven Spinola directed building owners and brokers elsewhere in Manhattan to hold rents at levels pertaining before the attacks as a specific prohibition against price gouging as displaced tenants sought accommodations during the fall of 2001.

The provision of Liberty Bond financing from the federal government provided important funds to areas affected by the 9/11 destruction, not only at Ground Zero but elsewhere in New York. Through a transition of mayoral administrations, and with a bipartisan effort on the part of New York's congressional delegation and the support of the private sector, funds to enhance the feasibility of real estate market improvements surely stabilized the property sector and then accelerated its recovery.

The New York City Central Labor Council established an Emergency Employment Clearinghouse to assist the 75,000 workers—across a wide spectrum of occupations and income levels—who lost their jobs in the aftermath of the attacks. As was the case in the fiscal crisis of the 1970s, many of the entities in New York's complex network of relationships set aside differences in taking a "high road strategy" that undoubtedly assisted the economy and the property markets.

That road was not without its potholes. In the national and local economic uncertainty of 2001–2002, which included a recession officially dated from March to November 2001, many Manhattan businesses released a supply of sublease space to the market, which elevated office vacancies and put downward pressure on rents. By the first quarter of 2002, Cushman & Wakefield was reporting a 7 percent decline in market rent, a situation that discouraged any further construction of market-rate offices in the short term. That extended a period of low development volume dating as far back as 1991 and set the stage for recovery in vacancy and rent.

By 2005, office vacancy in Midtown had returned to pre-9/11 levels (under 8 percent), rents were rising at a 5 percent annual pace, and absorption hit 4 million square feet in the second quarter of 2005. This was supported by employment growth that reduced the city's unemployment rate to 5.1 percent by August 2005, the lowest jobless rate since July 1988. Seven World Trade Center, started by Silverstein Properties in May 2002, was completed in 2006, commencing the return of business to the Ground Zero locations. Meanwhile, the evolution of Downtown Manhattan toward a 24/7 district was being accelerated by the conversion of older office buildings to residential use in Tribeca and along Wall Street.

Despite the beginnings of economic recovery in 2005–2006, the outlook for office market demand in Downtown was still considered to be uncertain during those critical years [when] the decisions to rebuild all 10 million square feet of office space at the World Trade Center were front and center in the local press. The remarkable optimism of Larry Silverstein predicted that Downtown's vital transportation network would be rebuilt and, together with new state-of-the-art buildings, would—and did—attract new and vital office tenants beyond the traditional firms in financial services.

In retrospect, New York, and its real estate market, has been well served by prior experience in dealing with crisis. Many of the key actors who helped shape the post-9/11 recovery had lived through the city's deep decline and economic restructuring a quarter-century earlier. Though shocked, they were propelled toward problem-solving rather than panic, to collaborative effort rather than narrow definitions of self-interest. Now, nearly two decades later, we are the beneficiaries.

Future of Lower Manhattan's Office Market

In the Civic Alliance report's[5] chapter on the office market, Hugh Kelly noted that the World Trade Center site, both in terms of its location and its critical hub for regional transportation, was best suited to provide the needed commercial space for an anticipated market recovery.

In the report's chapter on "Rebuilding Lower Manhattan for the Creative Age," Richard Florida considered that rebuilding with financial office towers would be unsatisfactory for changing trends in the regional and national economy. He proposed that a likely future of mixed-use development, though neither all high-tech nor tourism would emphasize a markedly improved transit system in Lower Manhattan. Florida's prediction for NYC was that it would transition to be an early pacesetter in the creative age.[6]

NYC's leading broker of office space, Mary Ann Tighe, then Regional Director of CB Richard Ellis (CBRE), noted the results of an October 2005 feasibility study. She surmised that:

> Timely development of the World Trade Center site will be necessary to meet anticipated office demand over the next ten years ... [and] more than 6.6 million square feet of older office space had been converted to residential development between 1997 and 2001. [Hence] the World Trade Center and its surrounding environment constitute an entirely new neighborhood, and ... the WTC site will be the most compelling relocation alternative for those tenants priced out of Midtown.[7]

Presenting the findings of her report at a breakfast meeting,[8] she listed five attractive features of Lower Manhattan that had emerged:

- Downtown offers some of the best values in the region with the incentives offered together with lower real estate taxes, providing a savings of 40 percent.
- Downtown's commercial tenant base is growing more diverse with the increasing presence of law, education, and health care.

- Downtown is 24/7 as a result of the substantial growth of residents in the area.
- New public spaces and amenities have been added, including the refurbishing and new building of 20 parks by 2007 and the opening of 15 museums covering a broad array of interests.
- Downtown has more transportation investment than any other area in NYC with the building of more than $9 billion in transportation improvements.

Tighe further noted that NYC's office employment was growing at approximately 1.4 percent and that the larger tenants, requiring 100,000 square feet or more, were seeing decreasing opportunities in the Midtown area. Lower Manhattan was starting from a stronger competitive position with its lower pre-9/11 office rents and inventory spanning 19 blocks. Summing up, she noted that this was definitely not the Lower Manhattan we once knew, but now had a resurgence in residents and a vibrant urban life.

At the T+15 panel on the residential sector on May 18, 2017, Chris Jones, the Head of Research at the Regional Plan Association (RPA), evaluated the upcoming balance of supply and demand for employment space in NYC. To accommodate its recent growth of 80,000 jobs per year, the city needed to locate some jobs in Lower Manhattan, and this required new building. He warned that merely converting existing Class B space into Class A space in Lower Manhattan would not be adequate since it would leave the affordable Class B sector undersupplied.

Design Issues

The Question of the Skyscraper

Given the notability, and eventual vulnerability, of the soaring towers of the original WTC, the consideration of building new skyscrapers opened many avenues of discussion in the months and years following 9/11.

First, for architects involved with the rapid building of skyscrapers throughout the world, the consideration of a new building provided an opportunity to explore alternatives for the skyscraper typology. Re-envisioning the form of the skyscraper, the functional activities contained within, and the placement of such a building in the urban landscape all posed challenges for architects, engineers, and urbanists.

Other aspects of the skyscraper were to be critically examined by engineers given the failures of structural performance in the WTC under attack. The height and structural feasibility had been under consideration as taller and taller buildings were proposed in the Middle East and Asia, but with excessive height came crowding out of much usable space at the ground and lower floors. This loss of rentable square footage to structural frames eroded the revenue potential while significantly increasing building costs in these super-tall buildings.

Also strongly highlighted by the collapse of the WTC was the question of occupant safety within skyscrapers, particularly with respect to evacuation. Admittedly, this instance of the buildings being subjected to the impact of a Boeing 767 jet and the burning of its 20,000 gallons of high-octane fuel was not typical, but the intended evacuation process for the towers exhibited inefficiencies and dangerous impediments. In the aftermath, not only structural improvements and resilience were being considered but also the logistics of evacuative formations were being re-evaluated.

The tragedy of 9/11 had additionally brought into focus the vulnerability of corporations and firms when so geographically concentrated. Of first consideration was the decentralization of data and technology, while adding inbuilt redundancy and backup. Though also of concern to many organizations whose assets were predominantly their professionals was the viability of the firm when such significant personnel losses occurred. Although functional integration had demonstrable corporate benefits, large financial firms such as Morgan Stanley later undertook strategic operational decentralization, and this mode of business has since been adopted by most financial services firms globally.

Rebuild the Towers!

Despite the cautionary aspects of skyscrapers learned from the WTC collapse, there emerged some very passionate and vocal groups in support of rebuilding the Twin Towers exactly as they had been. These groups included Build the Towers, Build a New WTC, and Team Twin Towers; they saw the reconstruction of the old form as symbolic of their defiance to the perpetrators of the tragedy. Additionally, there was significant belief in the positive economic impact of rebuilding the 10 million square feet of office space, though not all were convinced that a large commercial complex and the twin towers format were the most desirable option.

Opposing this emotional attachment to the twin-tower form, there was an opinion that to replicate that structure was not only less feasible, but also presented a much too poignant reminder to those whose loved ones had perished in the destruction. Paul Goldberger[9] reminds us:

> The problem we face, in short, is that if we express the tragedy of September 11 in its full magnitude, it is hard to imagine that we will be making a neighborhood that is as easy and comfortable to live and work in as we want it to be. But we cannot trivialize the events of September 11, and the lives lost, by turning the neighborhood into just a pleasant theme park, either. It is not a mall, and it is not a festival marketplace. It has to show some scars. There has to be some sense of the extraordinary.

Notes

1 Grubb & Ellis Special Report: *New York City's Office Market Post 9-11 Update* (December 2001).
2 From 2017–2022, Hugh Kelly served as a Special Advisor and Chair of the Curriculum Committee of Fordham University's Real Estate Institute, New York.
3 Hugh F. Kelly, *The New York Regional and Downtown Office Markets: History and Prospects After 9/11*, report prepared for the Civic Alliance to Rebuild Downtown New York, August 9, 2002, published in the *Economic Development Working Group Reports*, September 2002.
4 May 24, 2018.
5 *Economic Development Working Group Reports*. Prepared for the Civic Alliance to Rebuild Downtown New York, Alice Rivlin and Rosemary Scanlon, Co-Chairs. September 2002.
6 Richard Florida, "Rebuilding Lower Manhattan for the Creative Age: Implications for the Greater New York Region." Prepared for the Regional Plan Association and the Civic Alliance to Rebuild

Downtown New York, April 26, 2002. Published in *Economic Development Working Group Reports* (September 2002).

7 Mary Ann Tighe, *World Trade Feasibility Study*. Report for CB Richard Ellis (CBRE), October 21, 2005, p. 2.

8 Mary Ann Tighe, *Downtown Market Forecast Breakfast*, October 20, 2005.

9 Goldberger, *Up from Zero* (Random House, 2004), p. 250.

Chapter 9: Resolving Development Impasses

From the start, Larry Silverstein had been determined to rebuild the 10 million square feet of commercial space that he had leased from the Port Authority and lost on 9/11. Contractually, he was committed under the terms of the 99-year lease to fully restore the commercial space and generate revenues to meet his ground lease payments to the Port Authority. The major questions were:

1. Who would have logistical responsibility for the rebuilding?
2. How was the rebuilding to be financed?

Lynne Sagalyn described the central challenge: "Everyone understood that the lease agreement had to be negotiated."[1]

Rebuilding Responsibility

Although the replacement of the specific buildings—the office towers, the subterranean retail—were legally to be undertaken by the lessees, Silverstein Properties and Westfield, respectively, those structures were supported by foundations that incorporated the vast underground systems of subway lines and the PATH train terminus. Further, anticipated changes to land use on the site were needed to accommodate the proposed 9/11 Memorial & Museum and public open space. Therefore, for efficiency and centralization of the project management, the various parties to the leases and the legislative authorities agreed that the rebuilding of the WTC site, as a whole, was to be implemented through a new agreement between the Port Authority as landowner or lessor and the lessees of the buildings, Silverstein and Westfield. The LMDC oversaw the formulation of this new lease agreement.[2]

By December 1, 2003, the Restructuring Letter of Agreement directed the key actions to achieve the implementation of the new plan for the WTC site:

- There was an immediate "swap" of the footprints of the destroyed Twin Towers (which were recognized as hallowed ground destined to become the Memorial) in exchange for five surrounding sites to accommodate the rebuilding of 10 million square feet of commercially viable, Class A, above-grade, rentable office space in towers with at-grade lobbies.

 DOI: 10.1201/9781003476139-9

- A non-dimensioned but diagrammatically clear site layout was agreed by the parties, which effectively committed them to the new LMDC Master Plan.
- Principles for future negotiation were agreed upon, and the parties committed to cooperate fully with each other fairly, reasonably, and in good faith and to fine-tune and finalize the Master Plan sketch and renegotiate the leases and other agreements to reflect the new terms.
- Westfield's claim as the lessee of the 400,000 square feet of retail space was bought out by the Port Authority.

Over the following year, the parties attended to their respective commitments set forth in the agreement of December 1, 2003, and by Thanksgiving (late November) of 2004, had achieved further resolution of the physical layout of the site. Development sites for Silverstein's towers were now identified and fully dimensioned on the Master Plan. Additionally, during this time, the New York City Mayor's Office had become comfortable with the plan and the impact on the city and joined with the Port Authority and Silverstein as signatories to the Design and Site Plan Thanksgiving Agreement.

The restoration of the infrastructure then proceeded according to the Site Plan and Silverstein attended to the planning and financing of his new towers.

Economic Challenges

Although economist Hugh Kelly and real estate broker Mary Ann Tighe testified at a New York City Council meeting in March 2006 that rents would rise sharply as the real estate cycle improved, the delivery of 10 million square feet within a few years seemed to some to be a large volume of new inventory for the NYC office market to absorb. This market challenge was exacerbated by the general compulsion to rebuild the site all at once rather than in phases, although the LMDC had disagreed with this schedule in its 2004 report.

The NYC Mayoral Office suggested alternative developments of residential and hotel buildings on the 16-acre site, and at most required only three office towers, which would deliver 6 million square feet. Although the Port Authority Chair at that time, Anthony Coscia, agreed with the reduced amount of office space, the change of use of the site would require legislation by both NY and NJ state assemblies, with the latter now led by Governor Corzine.

In January 2006, Silverstein arrived at the proposition that he could build the 2 WTC, 3 WTC, and 4 WTC towers, and wished to commence. However, as Sagalyn reports, the opposition remained strong: "[There were] fierce editorials for the next six weeks. The *Wall Street Journal* editorial said: 'It's a disgrace (NY's dysfunctional political culture), … a case study in government incompetence and buck-passing.'"[3]

Meanwhile, the Insurance Battle Was Underway

At this time, the current estimates of the cost of rebuilding the 10 million square feet plus retail plus infrastructure ranged from $9 billion to $12 billion.[4] For his portion of the structures, Silverstein, together with his lawyers, Wachtell, Lipton, Rosen & Katz argued that the destruction was caused by two occurrences—the Twin Towers were hit by two planes, with the destruction of each tower contributing to the collapse of the other. The legal opinion of international experts was provided in support of this interpretation. Therefore, Silverstein Properties sought $7.1 billion in insurance funds, which represented twice the established replacement cost of the insurance agreement.

The insurance claim was further complicated by 20 entities that had combined to form the club of insurers. The destruction occurred just six weeks after Silverstein's acquisition, and as typical of the process, the transaction had closed with a "binder" that provided an outline of the coverage that was intended to be superseded by a more detailed, formal insurance policy document. During the period following the closing of the sale, approximately half of the insurance club members had resolved their commitment to the final document, with the rest still committed by the binder. As is also common practice, the language of the final document more precisely defined the terms, and the specific use of the word "occurrence" had been adopted. It was the discussion of the semantics of the term "occurrence," in contrast to the initial binder's use of the term "incident," that caused those insurers committed to the final document to be liable for twice the replacement amount—for example, the Traveler's Plan portion was based on total proceeds of $2.2 billion rather than the estimated building value of $1.1 billion.

Even with these additional revenues, however, given cost inflation and the new resilience and security requirements for the replacement buildings, Silverstein estimated he would not have enough capital to rebuild the entire 10 million square feet. Also, the Lower Manhattan economy and office market remained weak following the 2000–2001 recession.

Seeking clarity on the situation, in the fall of 2005, New York City's Deputy Mayor for Development, Dan Doctoroff, commissioned the investment bank Lehman Brothers to detail the difficulties of commercial development at Ground Zero. Also concerned with the economic potential of its land in the current circumstances, the Port Authority hired the office brokers Jones Lang LaSalle to assess the financial viability of developing the 16 acres of land. Both reports supported Silverstein's claim that he would not have the resources to rebuild the total 10 million square feet contained in the newly designed five towers.

April 16, 2006—Conceptual Framework of the Master Development Agreement

Therefore, it became clear that there would need to be a change in Silverstein's agreement with the Port Authority. As a basis for this, the investment bank Morgan Stanley was hired to determine the capital requirements and potential sources and prepare a capital plan for the project.

Governor Pataki demanded a speedy resolution of the impasse, and after an arduous process, two key principles of the revised agreement were resolved as:

1. The Freedom Tower (1 WTC) and 5 WTC (the "Deutsche Bank Tower") were to be developed by the Port Authority, and Silverstein was to build 2 WTC, 3 WTC, and 4 WTC.
2. The lease between Silverstein and the Port Authority on those three buildings would have a revised rent structure that reduced the financial obligations of the lessee in the early years following construction.

November 16, 2006—Finalizing the Master Development Agreement

While the structure of a financially feasible development program was being explored, an additional point of conflict between the Port Authority and Silverstein required resolution. This concerned the contractual cross-default provisions of the original 2001 purchase agreement, which is described by Lynne Sagalyn (2016):

> [T]hey actually signed separate lease agreements for each of the four towers Silverstein would manage and operate for the next 99 years. Under the cross-default provision of these leases, the group would have to perform on every lease; if one building became unprofitable to operate, it [Silverstein] could not just walk away without incurring a severe penalty because in doing so, it would lose all property rights on the other towers. ... Once the attack on 9/11 destroyed those towers, it destroyed the intended functionality of that contractual provision of the lease.[5]

The process, timing, and risks of constructing the new towers, however, made such an obligation much more difficult for Silverstein to meet, though NYC and the Port Authority did want to ensure an adequate performance on an economically critical mass of rebuilding. After substantial negotiation on the details of performance responsibilities, a Master Development Agreement (MDA) was executed in November 2006. Importantly, it established:

- the division of roles and responsibilities relating to the allocations of development rights on the Master Plan, and the effective relaxing of the cross-default provisions for Silverstein;
- further details of the revised rent structure;
- revised leases between the parties.

With this understanding among all parties, the project could then proceed.

Reflections of Janno Lieber[6]

I have often been asked, where did we guys at Silverstein Properties muster up the optimism to envision 10 million square feet of office space when the world thought the whole economy of Downtown Manhattan was in doubt. Where did the optimism come from?

Well, everyone who knows Larry Silverstein knows that Larry has the gland that produces whatever the substance is that creates optimism in human beings, starting with himself. So, a lot of our optimism was just Larry's view of the world. Larry fundamentally is relentlessly optimistic, even under the most difficult circumstances. And yes, that's a lot of it.

But there are other factors as well. Larry and our organization were looking at the role that we had to play. So the first thing that Larry did was say, "What can I contribute to this picture?" And what Larry had to offer was the ability to get an office building up, to persuade people that Downtown had a future as a vibrant commercial district, and to counteract the pessimism, the gloom, and the sense of uncertainty that surrounded Lower Manhattan in the wake of 9/11.

So under Larry's leadership, and with SOM in the lead as the professional architect, our organization got to work right away planning to rebuild 7 WTC. And this building now is over 10 years old because it was started so quickly. And in the end, it has been so successful.

A lot of folks were expecting 7 WTC to be a white elephant, including Mayor Bloomberg in those early days, although Bloomberg later became a great supporter of what all of us were trying to accomplish. But, lo and behold, 7 WTC was successfully leased to a really interesting mix of companies. And it proved that people would want to come back to a new, more vibrant, more pedestrian-friendly Downtown.

So, a lot of it was Larry's view and our team's view of real estate fundamentals. I remember Larry saying to me, "Well, I don't know how I'm gonna get from A to Z, but we have to go from A to B, and B to C. But one thing we do know is that no one is creating a lot of new office building sites with the best mass transit in New York City, so let's get started figuring out how we can make a vibrant business district for the city."

It was that simple. It was basic real estate fundamentals. The inventory of office space in Manhattan was dramatically aging. And that was evident even at the time of 9/11. More than half of Manhattan's office buildings were already more than 50 years old. We also knew that there was going to be a need for a new generation of office space, as the city was becoming less about manufacturing and more about the production of value and ideas through what takes place in office buildings.

There was also a growing, huge gap between the value of Downtown office space and Midtown office space, which Larry viewed as an opportunity. Downtown was being undervalued so that, in effect, you could create some momentum by creating great new space that would command greater values. All the analysis of trophy buildings that goes on in New York has established that new office buildings do have greater value. And that, in the end, was part of the optimism.

The final opportunity was mass transit. I've never traveled with Larry Silverstein on the subway. (There's always something to look forward to.) But Larry's a New Yorker, and I had a background in transportation, and so all of us at Silverstein Properties were obsessed with the fact that this was the one neighborhood that had 13 subway stations, a connection to New Jersey, and proximity to Brooklyn. As much as Midtown is our dominant business district, if you enter by the Port Authority Bus Terminal, or Penn Station, or Grand Central, or even the Times Square subway station, you may be 20 minutes from your desk.

Whereas in Lower Manhattan, because it's a narrow island with so much subway service, we always believed passionately that this was going to be the best place to work in a 21st-century context—at a time when more people wanted to live in the city, take mass transit, and the suburbanization trend was being reversed.

So, for all those reasons, there was a lot of optimism that fueled our team in addition to the chemistry of Larry Silverstein, the man.

A Further Impasse in 2010

While the Port Authority proceeded with the development of 1 WTC, engaging the Durst Corporation as partners, Silverstein worked on the plans and funding for his three towers, 2 WTC, 3 WTC, and 4 WTC. However, the financial crisis of 2007–2008 adversely affected the office markets in NYC and the feasibility of the new towers was at risk.

Also, during 2009, there had been continual back-and-forth disagreements over a number of remaining issues between the Port Authority and Silverstein:

- Silverstein's development and management fees, as stipulated in the 2006 MDA, had, by 2009, totaled $177 million. These were being drawn from the insurance proceeds that were estimated by *The New York Times* in the prior year as having only $1 billion remaining from the $4.5 billion received.
- Silverstein continued to be obligated to pay the Port Authority $120 million per year in ground lease rent under the terms of his initial purchase.
- Silverstein levied fines on the Port Authority for being late on delivering the necessary infrastructure on the WTC site.

Responding to the deterioration in market conditions, and now under the new Executive Director, Chris Ward, the Port Authority sought an updated view of the market demand with respect to the schedule for rebuilding 2 WTC, 3 WTC, and 4 WTC. The Port Authority hired Cushman & Wakefield (C&W) to estimate anticipated demand for office space in the coming years. C&W concluded that if all three towers were built by Silverstein at once, the additional inventory of office space would have a "potentially destabilizing effect" on the NYC office market.[7]

Taking this assessment into account but also pressing for the commencement of 4 WTC by Silverstein as soon as possible, the Port Authority offered to "backstop" the revenues on that building by taking a master lease on the 2 million square feet. However, the amount of rent to be paid on this master lease became a point of contention between the parties.

An impasse persisted despite efforts by Mayor Bloomberg, the NYS authorities, and public criticism of both sides.

Amidst the stalemate, Chris Ward suggested a Plan B: The Port Authority would move ahead with building the infrastructure for the Transit Hub, to be renamed the Oculus, and "walling off" the planned mechanicals for the base of 3 WTC. Additionally, at this point,

Westfield returned to the project and offered to build a $1.3 billion retail complex integrated within the Oculus.

Reflections of Chris Ward[8]

The history of the development of the World Trade Center will be written numerous times; it is by definition a site without a single narrative. Over time, this narrative will only likely increase, as the world continually grapples with the growing political interpretation and role of terrorism. From the beginning, this overlapping narrative was both its guiding vision and its greatest liability. The complete destruction of 16 acres on the island of Manhattan, the loss of an iconic skyline, and, most importantly, the deaths of almost 3,000 people by a foreign force were certain to drive the fundamental question of what defines the city and the nation as a whole. Interwoven through the meta-narrative was the uniquely New York response to the catastrophic loss of first responders—the firefighters and police; to the stories of the day-to-day lives of both those lost and those who survived; and, finally, to each New Yorker's sense of their city and their own self. It was a hole in the heart of the city and the nation, yearning to be filled.

The quest to fill the site with meaning began almost immediately. First, the families of those who were lost pressed hard for memorialization, but with the numbers came many visions. Some sought complete sanctity, others looking to move on pressed toward a robust rebuild, oftentimes in the rhetoric of sending a message to the terrorists. The real estate industry, led by the World Trade Center leaseholder Larry Silverstein, similarly saw the loss of the Twin Towers as a threat to the economic future of the city. Not unexpectedly, the government saw itself as the key driver in the response: The Port Authority of New York and New Jersey still owned the site, the city controlled the development and regulatory environment, and both states oversaw the Port's governance. Finally, the civic community saw a role in guiding redevelopment as an opportunity to correct historic flaws in the original design and create a new model for a reimagined Downtown. The interplay of these different goals and objectives was never static; at times, each took on meaning and control of the site. But throughout, the families remained a persistent constant.

Leadership soon became the essential question: Who would actually lead the rebuilding project. Immediately after the attack, the Port Authority began an interim restoration plan. As I was then the Director of External Affairs, we recognized that a comprehensive response—albeit temporary—was critical. A plan was developed. It included a temporary museum on Pier One on the Brooklyn waterfront. The idea was that walking from the site over the Brooklyn Bridge and looking back at the vacant skyline would capture some sense of the loss. It (the temporary museum) would also house certain artifacts of the day. The site was also to be draped as commemoration and a visual barrier; the damaged street grid needed to be greened and turned over to pedestrians. Politics intervened. On the cusp of a critical mayoral election, a city/state deal was struck to create a whole new entity to guide the project, blocking the leading Democratic candidate from a role. It was called the Lower Manhattan Development Corporation (LMDC). It was entirely under the control of Governor Pataki.

But the process to fill the void had already begun. In an unprecedented public forum, the civic advocates convened a visioning session for all comers at the Javits Convention Center. Thousands of people came with ideas and schemes for what should replace the Twin Towers. The city poured its heart out. But the visions needed a structure, and the LMDC took the lead in pulling ideas together but also in setting clear goals and a framework for what would come next. Another more limited convening took place when the LMDC called the leading architecture firms to give their vision of the future. By now, I had left the Port Authority and joined the Bloomberg administration as Commissioner of the Department of Environmental Protection, but I remember how unimpressed I was with the assorted schemes; some were, to my mind, almost laughably out of touch with the urban fabric of New York, no matter what the future might be.

Two schemes competed right to the very end, when Governor Pataki chose Daniel Libeskind's over Viñoly's THINK plan. At first, people thought the plans were designs, what would actually get built, but it was a Master Plan that was selected. The actual architecture and components of the site still needed to be drawn. And soon, it was clear that the site would not be defined by one idea, not even two; there was simply too much demand for more, almost more of everything. In short, there would be a memorial, a museum, a whole new transit hub, a re-established street grid, and more than 8 million square feet of new real estate in four new buildings, anchored by a new tower dubbed the "Freedom Tower," rising to a symbolic peak of 1,776 feet. A simple but elegant design by David Childs at SOM [Skidmore, Owings & Merrill] was well received by the public. Another public selection process had ensued for the Memorial Plaza. Initially designed by Michael Arad, it went through a series of reiterations until finally complete, but not without controversy and drama. Perhaps the most iconic structure was to be the new Transit Hub designed by Santiago Calatrava. Evoking the wings of a dove, with its anthropomorphic curved design, the gleaming white structure would stand in contrast to the right angles of the city. This design would later bedevil the construction contractors and its budget.

Like much of the history of the City of New York, the city got its plan, a monumental one, a plan that sought to answer every question and provide every answer to the meaning of the site. In a sense, it was to be a whole new city within the city, for it had everything a city aspires to: public space, culture, real estate, a museum to history, a new style of street grid, and soaring architecture. It was all there in monumental fashion. Clearly, politics and power had played a role, a key role, but given New York City, it was almost inevitable that such a monumental vision would shape the site.

But the many components of the plan also paralyzed the site. When I became Executive Director in 2008, the site looked largely unchanged. It was defined by nearly everyone as a "pit," derided by the tabloids and completely stalled in the public's mind. Frustrations were reaching a fevered pitch, with the federal government threatening to take over, the city and state fighting constantly, and the endless symbolic groundbreakings had generated a cynical boiling point with the public. Most importantly, no one believed in the schedules and budgets, and yet these had remained virtually unchanged despite the transparent contradiction of the site. Newly installed, Governor David Paterson of New York directed me to do two things: Reset the schedule and create an actual budget. This was to be done in two months, but soon it was clear that was an impossibility; we would need four months to really get things back on track. Tactically, we decided to get there in two steps. The first, mostly for credibility, was completed in two months. This was intended to make clear to the public just how complicated a project it was and buy us time. Two months later we released the road map for moving forward. It began with a clear statement of the obvious: The schedules and budgets were completely out of line and a whole new set of deadlines had to be established. Our candor was met with grudging acceptance, and the narrative was, "At least someone is finally telling the truth."

As it turned out, meeting this four-month challenge was the easy part. Now the real challenge was how to actually build it. This was not a simple construction question. For as we described it, the site was like the children's game Pick-Up Sticks. Every single part of the project touched another part; the foundation of 4 WTC was the curtain wall for the below-grade Museum; the heating and cooling system for the entire site ran through 2 WTC and 3 WTC; the new PATH train snaked through the entire site. Nothing was independent from anything else. As a result, all the stakeholders and the vested interests had an equal seat at the table. The monumental design had no inherent sense of priority, hence the endless gridlock and delay. We struggled with a way to redefine the goals and expectations to break the logjam.

It was soon clear that the approaching 10-year anniversary date was critical, that dates mattered, and that not to deliver a public plaza was just unacceptable. How could President Obama and former President Bush visit an unfinished site? How could the families gather after 10 years and feel no sense of closure? How could the city remember when so much was left still undone? The problem was that even our aggressive schedule was showing the public plaza completed only by 2013. This was not going to work. Challenging the Port Authority's Engineering Department, long nights were spent tearing the plans apart and working to find a solution. Finally, turning the world on its head, they proposed building the plaza from the top down, the ceiling of the Museum would be the floor of the plaza, counter to the logic of starting at the bottom and working your way up. Numerous reviews confirmed the constructability, but an incremental cost to complete the interstitial space would be incurred, around $60 million. We deemed it a worthwhile investment, not just because it would deliver the plaza date certain, but

9.1

Announcing the resolution of negotiations. Left to right: Michael Arad (9/11 Memorial architect), Joe Daniels (9/11 Museum CEO), NY State Assembly leader Sheldon Silver, Mayor Mike Bloomberg, Port Authority Executive Director Chris Ward, architect Daniel Libeskind, developer Larry Silverstein.
Source: Joe Woolhead, courtesy of Silverstein Properties.

almost more importantly, it would create a new ability to set priorities and address the conflicting interests. No one could argue with the delivery of the plaza; in a sense, it was the sacred heart of the site.

Now we could address the problem of monumentalism because we turned the site into a construction job. This was a turning point in the project, making all the difference. Construction and budget problems would remain, particularly the challenge of the Calatrava Transit Hub. His vision, as soaring as it was, presented a unique set of problems as everything was essentially bespoke. With no right angles, each piece of steel required its own casting, driving up costs and lengthening schedules. One issue loomed largest: the notion of a column-free space, that somehow all the steel beams could be placed and rested without any delay. It was clear that this would prove impossible and that column support was essential. To his credit, Calatrava did not balk entirely at our suggestion, although it took a while to fully convince him of the importance. But the columns were added to the design, eliminating significant construction risk.

The benefits of breaking through the monumentalism and delivering the plaza by the 10-year anniversary had other values as well. The Port Authority was locked into intense real estate negotiation with the leaseholder, Larry Silverstein, which ended up in arbitration. Silverstein was seeking billions in lost real estate value, while we argued that he never had the resources to meet the market regardless. Most importantly, we argued that the site was now under control and the construction schedule was being met. The arbitration decision acknowledged the progress, and provided no damages payment, and redirected the negotiations to recommence. This was a huge win for the Port Authority. Subsequently, partial financing was agreed for 4 WTC and 3 WTC, while deferring any public funding for 2 WTC.

As I look back, the World Trade Center site as it is today was born out of a singular event in U.S. history, one that touched numerous constituencies and priorities, but creating a host of competing political interests. Its early years reflected all the emotions and visions the site could possibly endure, and with them came delay and cost overruns. Monumentalism had washed over the 16 acres, blurring the hard reality of how to actually build such a vision. With Governor Paterson's leadership and political protection, we were able to create a credible framework for rebuilding that the public and elected officials accepted. Delivering the Memorial Plaza by year 10 was the rock that broke the logjam, allowing us to both simplify the goals and set a clear schedule going forward.

So, on another clear, sunny, cloudless September morning, President Obama and former President Bush, along with Governors Pataki and Christie and Mayor Bloomberg, all gathered with the families on September 11, 2011 at a completed Memorial, with the water flowing deep into the voids.

March 25, 2010—Restructuring Framework

In further resolution of the 2009 impasse between Silverstein and the Port Authority, on January 26, 2010, an arbitration panel released its decision regarding the dispute: The Port Authority was not required to pay fines to Silverstein for the delays in completing the infrastructure on the site, though the lessee continued with his rental obligations.

Though this settled the most significant disagreement, a number of other issues remained unresolved that required resolution by arbitration within 45 days. Sagalyn[9] lists these issues as:

- market risk;
- rent payments;
- insurance dollars;
- financial support or "backstop";
- development fees;
- Silverstein's "skin in the game."

In these 45-day deliberations, Tim Gilchrist represented the State of New York, Bob Lieber, as Deputy Mayor, represented NYC assisted by Seth Pinsky of NYC Economic Development Corporation (EDC), while Wachtell, Lipton, Rosen & Katz represented Silverstein Properties.

> Cultural, ideological, and personal divides further colored the negotiation environment. The people working on the private side of development and their

> counterparts in the public sector operated in distinctly different worlds. Timelines and tenets for making decisions differed, and an understanding of how to view the transaction from the other's perspective seemed to be missing.[10]

Important to this phase of negotiating an agreement that would enable the project to move forward was the new presence of a person who represented a public authority, but was also an experienced dealmaker in private real estate activities.

Robert ("Bob") C. Lieber had been the head of the NYC Economic Development Corporation, but when Deputy Mayor Doctoroff stepped down at the start of 2008, Lieber succeeded him. He came to public service after decades in the private sector, most recently as a global head of real estate investment banking and responsible for a private-equity fund at Lehman Brothers. While in that earlier role in 2005, he had undertaken an advisory financial analysis for the NYC Mayor's Office of Silverstein's capability in undertaking the planned rebuilding of 10 million square feet of office space.[11]

With a focus on achieving a deal between all parties while meeting the policy imperatives of the city, Lieber also understood Silverstein's financial decision-making—its objectives and its processes—and was able to encourage the developer to move forward on new decisions and put the past disagreements between the parties, particularly with the Port Authority, behind him. Sagalyn reports Lieber insisting: "This is no time for laments. You have to absolve yourself of this versus seeking revenge."[12]

Tim Gilchrist Reflects on the Final Stages of This Arbitration Process

In March 2008, Governor Spitzer resigned, and Governor Paterson assumed office. I became the Governor's lead person on working with LMDC, the PA, and Silverstein negotiating the 2010 post-arbitration agreement on 2 WTC, 3 WTC, and 4 WTC, the demolition of the Deutsche Bank Building, the future of the proposed cultural center, and the ongoing coordination among the state and city agencies.

Once the arbitrators issued a decision in January 2010, Silverstein Properties and the Port Authority renewed negotiations on a new framework for the WTC site. In that the Governor and the Mayor wanted a resolution to the conflict and a resumption of construction, I was assigned, as was Deputy Mayor Bob Lieber, to join the negotiation with the PA and Silverstein Properties.

It soon became clear that the long-standing issues and conflicts between the PA and Silverstein Properties would require intense intervention on both Bob's and my part and that innovative ways of protecting the PA balance sheet and supporting Silverstein's finances would be required. I basically relocated to NYC for February and March of 2010 to work on the project and coordinate daily with Larry Schwartz, Governor Paterson's Secretary.

We relied on the PA staff and NYCEDC staff for financial analysis in the negotiations. I worked primarily with the Port Authority, while Bob worked with Silverstein. We, at times, acted as middlemen, discussing issues separately with each party and looking for common ground. Other times we attempted to clarify and narrow the issues to settle them with both parties, protecting core needs, but finding a way forward.

One of the major issues we needed to resolve was that by the very nature of the Port Authority structure versus the commercial structure of Silverstein, both looked at risk and debt in different ways. The PA had generated a large amount of debt and had invested major resources into the site. Silverstein, as any developer would, limited their own equity investment and risk to their balance sheet. Neither was wrong—they just approached the issue from different perspectives. Bob and I worked to find common ground and to make it clear to both parties we needed to find a way to success. The bi-state nature of the Port Authority and the involvement of the New Jersey Board members also provided another level of complication.

In the end, the city and state, and both the Port Authority and Silverstein found innovative ways to provide backstops in case the buildings were not fully occupied when completed, and the deal was completed in late March 2010. Shortly after, I left the Governor's Office to head up the Moynihan Station project.

October 11, 2010—Amended and Restated Master Development Agreement

Once the issues were resolved in principle within the mandated time period, they were drafted into an Amended and Restated Master Development Agreement that provided the framework for continued rebuilding of the WTC site. Finalized on October 11, 2010, the major resolutions of this agreement were as follows:

1. The Port Authority asked NYC to give up the scheduled payments in lieu of property taxes (Pilot), which had been originally required by the municipality for the WTC site and offered a fixed amount that provided necessary expense relief during the rebuilding and early lease-up years of the project. With his background as a leading financier, Bob Lieber ran an intense negotiation to arrive at the compensation amount of $130 million over 15 years.
2. NYS was asked to contribute $200 million, comprised of $80 million in capital and $120 million through the guarantee of lease backstops on certain portions of the property. However, this was conditional on Silverstein's obtaining an anchor tenant (based on the perception that the office-space markets had recovered) and private lender financing for 3 WTC in addition to contributing further equity of $300 million.
3. Regarding the development and management fees being earned by Silverstein, while these dues were to be paid on 4 WTC, no further fees were to be paid by the Port Authority on 2 WTC, and the terms for 3 WTC were structured in more detail.

More constraints were added to Silverstein's allocation of the insurance proceeds, with the Port Authority achieving a priority for payment on the 4 WTC and 3 WTC towers. To alleviate some financial burden on Silverstein, the Port Authority abated the payment of

9.2
Amending the Master Development Agreement. Left to right: Janno Lieber, Robert C. Lieber, Larry Silverstein.
Source: Joe Woolhead, courtesy of Silverstein Properties.

ground rent for 2 WTC and 3 WTC. This resolution enabled the infrastructure of the site to continue, and also provided clarity on the schedule for the office buildings.

With this new approach, Bob Lieber for NYC and Tim Gilchrist for NYS were able to achieve the 2010 Agreement. Sagalyn[13] concludes:

> [It was] a larger political story, a story of how public officials marshaled scarce public resources and reluctant private funds under great pressure to advance the ambitions of rebuilding at Ground Zero, and how their respective senior staffs moved mutually suspicious adversaries toward an overall agreement. The story would illustrate how financial risk was negotiated and shared, if not equally, in ways designed to assure that each of the four principals had "skin in the game." It would illuminate the art of political deal-making.

2011 Challenges

Scott Rechler was appointed as Vice Chairman of the Port Authority's Board of Commissioners in June 2011 and served until October 2016. In this role, he was in charge of the Port Authority's overall capital planning, $30 billion in total, including the rebuilding of the WTC. He had to view the WTC rebuilding within this context: It required that the economics of the rebuilding of the site would be subject to the needs of the Port Authority with respect to all its port, aviation, and transit services. In this, the assignment was substantially different from a typical institutional capital project.

But it was undoubtedly also a high-profile assignment. In discussion with the authors, Rechler recalls:

> When I was appointed Vice Chair, the Governor said, "Get the WTC done. The world is watching!" The rebuilding of One World Trade Center by Larry Silverstein had restarted after the October 2010 Agreement, but the more emotionally charged projects, the Memorial and the Museum, had lost momentum. Particularly the Museum project had become the "battleground" for the broader conflicts of the overall WTC rebuilding. We also needed to balance the sensitivities of the families and the remembrance of the tragedy, especially with respect to the Memorial and the Museum.

The issues were politically complex and there was little room for compromise. Following the resolution of the 2010 Agreement, all parties had been scarred. Fortunately, however, Rechler had substantial and solid relationships with many of the participants, and he was not involved in past conflicts on the site. He realized the need to revive the overarching administration of the project's components:

> When I got there in 2011, there was no budget, no timeline for the Museum, the Memorial, the Performing Arts Center (PAC), or the Greek Church. At that time, things were so bad that I gave the PAC a 1 percent chance of ever happening. Yet, this morning, 12 years later, I was at a breakfast meeting with Mayor Bloomberg and the Advisory Board for the PAC, and it is under construction and almost fully funded.[14]

Rechler's approach was to aim for accountability and transparency. He took an apolitical approach, emphasizing the basics. "Let passions cool..." was his approach. Then, he restarted the process by focusing on the Museum project.

To reorganize the Museum project, Scott established and chaired a specific subcommittee of the board and set up monthly meetings, which he co-chaired with Dan Tishman (of Tishman Construction), Steve Plate of the Port Authority, Bob Steele (by then, Deputy Mayor of Economic Development for NYC), and others to work through the issues, set timelines, emphasize solutions. "The monthly meetings were a valuable tool," noted Rechler.

The new administrative activities worked closely in line with the resolution achieved in the 2010 Agreement and the implications for the planning, schedule, and timing of the Museum project. Additionally, the project was audited to show details of costs. However, he also concedes, in discussion with the authors, that the unexpected and disguised elements of nature assisted:

> At times we were spared by fate. While being delayed on the construction of the Museum, we had Superstorm Sandy, which flooded the WTC site and left 20 feet of seawater on the Museum's floor. If we had been farther along in the process, the artifacts would have been installed. We were lucky. The storm also showed us

where there were vulnerable areas in the infrastructure that had to be plugged up against future storms.

Timing Was Imperative

An important aspect driving the rebuilding of the WTC site was the imminent 10-year anniversary of the 9/11/2001 event. This, and future anniversaries, effectively set time-frames to drive the scheduling of specific actions for the years ahead.

In 2014, many issues swirled around the outlook for 3 WTC. First, the city's economy was just beginning to recover from the adverse effects of the Global Financial Crisis. Silverstein could not proceed with his development of the skyscraper without a major tenant to underwrite the financial commitment of lenders. An alternative was to have the financial backing, or guarantee, of the Port Authority itself.

Exploring the market further, Scott Rechler brought in Mary Ann Tighe and Peter Rigardi of CBRE for their views on the outlook for the office market.

After weighing the options, Rechler proposed that the Port Authority guarantee the financing of the 3 WTC project, enabling it to move forward. With his understanding of real estate financing, he also brought in Warren Buffett's firm, Berkshire Hathaway, to advise on the potential issuance of bonds to fund the project. Eventually, Goldman Sachs agreed to the assessment of the strong bond market and proposed underwriting the bond issuance. The project proceeded, attracting the technology company, Group M, as an anchor tenant, and Rechler described it as "a major success for the WTC site."

Other Stalled Rebuilding

Another challenge for Scott Rechler in his new role was the standstill in rebuilding the St. Nicholas Greek Orthodox Church. As this was near the planned Transit Hub, it would be impacted by the design and construction process of that substantial infrastructure. Once again, his solution was to commence a process of regular meetings to resolve the issues, one at a time.

Commercially important at this time for the Port Authority, and the rebuilding project generally, was the retail plan for the Oculus at the World Trade Center (the new Santiago Calatrava-designed Transit Hub) and the large shopping mall by Westfield. The Port Authority had anticipated receiving rental revenue based on forecasts of future retail activity, though the outlook was seeming less certain. As an alternative, Scott Rechler proposed that the Port Authority would capitalize that future revenue in a sale of the retail space to Westfield—which it did, for $3.2 billion—providing a much-needed cash infusion for the Port Authority.

Notes

1 Lynne B. Sagalyn, *Power at Ground Zero: Politics, Money and the Remaking of Lower Manhattan* (Oxford University Press, 2016), p. 389.
2 This negotiation was overseen by the LMDC President at that time, Kevin Rampe, who applied his legal and legislative expertise and experience.
3 Sagalyn, *Power at Ground Zero* (2016), p. 440.
4 Sagalyn, *Power at Ground Zero* (2016), p. 440.

5 Sagalyn, *Power at Ground Zero* (2016), pp. 425–426.

6 Janno Lieber was then President of World Trade Center Properties LLC. In January 2022, he was appointed as Chair and CEO of the Metropolitan Transportation Authority (MTA).

7 Sagalyn, *Power at Ground Zero* (2016), p. 597.

8 Chris Ward was the Executive Director of the Port Authority from 2008 to 2011.

9 Sagalyn, *Power at Ground Zero* (2016), pp. 643–655.

10 Sagalyn, *Power at Ground Zero* (2016), p. 630.

11 Sagalyn, *Power at Ground Zero* (2016), p. 631.

12 Sagalyn, *Power at Ground Zero* (2016), pp. 631–632.

13 Sagalyn, *Power at Ground Zero* (2016), p. 632.

14 In Scott Rechler's discussion with the co-authors.

Chapter 10: Later Completions at the World Trade Center Rebuilding

The Oculus

The white-winged Oculus is an organic form at the center of the towers, Museum, and Memorial Pools. The structure's orientation serves as a lasting reminder of the attacks of September 11, 2001: Its central skylight is aligned with the angle of the sun's rays on each September 11, from 8:46 a.m. when the first plane struck until 10:28 a.m. when the second tower collapsed, and the beam of light washes across the Oculus floor.

Designed by the Spanish architect Santiago Calatrava, known for his soaring infrastructure, the functional purpose of the Oculus was as a hub for the numerous modes of public transit serving this area of Lower Manhattan. This was to incorporate the rebuilding of the destroyed sections of subway lines and stations, and the terminal of the PATH train, as well as provide covered walkways across the WTC.

As a project that began just one year after the tragic event and was completed 14 years later in February 2016, the Oculus not only serves as a physical connector for underground trains, but as Calatrava intended, "a witness of belief that we can overcome this tragedy, a symbol of the camaraderie of the American people, and a gift given to the community."[1] He continues:

> A new building should deliver a feeling of hope. And I have to say, it was not difficult. In the 14 years it took to create this building and the 14 years I lived in New York, day by day, following the work and the site, it was an extraordinary time. I was just learning from New York to do something extraordinary. What did I learn from the city? That New York landmarks are civic buildings. You think of the Brooklyn Bridge, for example, you think of the 5th Avenue New York Public Library—all of that inspired me. The entire city inspired me. I wanted to enter in this context. I wanted to emulate, not to imitate, but to emulate, Grand Central Terminal, such a civic monument and such a beautiful place.

Frank Sciame Reflects

At the time they announced the Oculus it was exactly what we needed down here. This was something special. Santiago (Calatrava) had a great reputation for

DOI: 10.1201/9781003476139-10

10.1
The Oculus.
Source: Joe Woolhead, courtesy of Silverstein Properties.

designing great transportation buildings. He provided the credibility needed to ensure that something great was happening at Ground Zero. I applaud Governor Pataki, Mayor Bloomberg, and the Port Authority's bold initiative to create an iconic PATH Station.

And it's a shame that it's getting such a tough rap on its cost because the Oculus itself cost only around $400 million. The rest of the $4 billion was spent below-grade, a great deal of which was spent on things like concrete-hardening and security. I have worked with Santiago—he is extremely cost-conscious.

When one visits the completed Oculus and observes the visitor's reaction—it is priceless.

Liberty Park

Carla Bonacci[2] organized and led design and project teams on planning for the WTC Liberty Park, including the landscaping and the WTC campus streetscape.

Carla Bonacci Reflects

I also credit the Master Plan process for establishing WTC Liberty Park above the Vehicular Security Center. We envisioned Liberty Park as a bridge to connect east and west, which it does (and beautifully), to the South Bridge that leads into Brookfield Place.

10.2
Opening performance inside the Oculus.
Source: Joe Woolhead, courtesy of Silverstein Properties.

Liberty Park is for the Downtown community. We worked closely with the community to develop this park. The residents wanted another place, other than the Memorial, where they could be separate. They wanted something that was quite different that would be their own, a quiet space where they could reflect. We worked with Michael Arad to create that nice frontage to the Memorial, the "Green Wall" which is the largest green wall in the country. Liberty Park, with its Green Wall, is very different from the Memorial, yet respects the Memorial it overlooks.

The Port Authority also had to decide on the issue of the famous cast bronze sculpture by German artist Fritz Koenig called *The Sphere*, which was originally on the plaza of the World Trade Center. The 25-foot sculpture was heavily damaged—but not destroyed—in the terrorist attack of 9/11. Some of the families requested that it be located on the Memorial Plaza, but eventually, the sculpture was placed in the raised Liberty Park at the south end of the site.

10.3

The Glade in Liberty Park.

Source: Joe Woolhead, courtesy of Silverstein Properties.

10.4

***The Sphere* amongst the ruins.**

Source: Joe Woolhead, courtesy of Silverstein Properties.

10.5
***The Sphere* again located at the WTC site.**
Source: Joe Woolhead, courtesy of Silverstein Properties.

Notes

1 Santiago Calatrava, *Oculus: World Trade Center Transportation Hub, New York* (Assouline Publishing, 2017), p. 16.

2 Carla Bonacci, AIA, PP, had more than 30 years of continuous involvement as a planner, architect, and builder at the WTC. From 2002 to 2022, she was Assistant Director of WTC Construction at the Port Authority.

Chapter 11: Important Partnerships

The rebuilding of Lower Manhattan was particularly challenging because of the multiple agencies, departments, and authorities involved throughout the three levels of government: federal, state, and city. Unlike in smaller cities, which are largely managed by the municipality, NYC includes the NYS's administration in its governance of major infrastructure, such as the subway and bus systems and certain roadways. Further, the major funding source for the rebuilding was the federal government which channeled the resources through various federal agencies with mandates that often differed from that of local authorities. Additionally, the public sentiment regarding the loss of lives and the need for increased security and anti-terrorism measures introduced new objectives into the process.

Reflections of Daniel Creighton[1]

Lower Manhattan Recovery Office (LMRO)

Although not brought into the project until 2005, I was honored to be involved in this high-profile project of such national significance. As a native New Yorker, I could not foresee a more important project to be part of. Upon arrival, I realized that the WTC site program faced significant hurdles to [be] overcome, with various visions, high political pressure, and competing objectives. Despite these challenges, everyone approached the work with a single vision: to complete the World Trade Center complex in a way that would honor the memories of the victims of that horrific day.

Significant federal funding had been granted to the area ($4.55 billion) not just for transportation recovery, but also for the economic revitalization of Lower Manhattan. Although LMRO requirements were not as stringent as regular federal program funds, they were still daunting, particularly when trying to move programs of this size forward in a seamless manner. My goal was to help navigate grantees throughout the process, not as a hindrance to the project, but rather, as a resource to help move the project forward. It was important for me to be an active participant rather than a sideline observer. It is my personal belief that a construction program of this size, with its multiple projects and multiple stakeholders,

DOI: 10.1201/9781003476139-11

requires us to coordinate the intricate dance between players that would never have worked without everyone choreographing every activity.

The recovery of Lower Manhattan after 9/11 involved so much more than just the World Trade Center site. With the worry of losing financial industries in the aftermath of the attacks, federal funds were utilized not only to rebuild but also to economically enhance the area to ensure that it was kept as the financial center of the world while also enhancing a livable community and tourism destination. Other projects besides the WTC Transit Hub were funded, such as the South Ferry Station on the #1 Subway line, Fulton Street Transit Center, the reconstruction of Route 9A between Chambers and West Thames streets, as well as various ferry and security projects. I truly believe these funds helped not only recover the financial industries in Lower Manhattan but also made the area a vibrant, livable community that has transformed significantly since 9/11.

Thinking back on the time spent on these projects, what always impressed me the most was the incredible dedication of every participant. Although there were constant disagreements, it was always out of a true sense of wanting to do what was right. Many of those who worked on the projects were themselves survivors of the attacks; thus, they took personal ownership in making sure this job was a success. Disagreements, more often than not, were conflicting priorities. In many situations, one project would have to suffer or be delayed so that another could move forward unhindered. The Lower Manhattan Construction Command Center (LMCCC) helped play a role when conflicts needed an arbiter to help move projects forward. In the end, priorities were clear, and personal interests always subsided for overall project goals.

From an oversight perspective, these disagreements could be troubling at times since advancements on one project would often lead to delays on another. FTA (Federal Transportation Administration) implemented a significant risk assessment program on the LMRO Projects to account for the unknowns that were likely to occur on these projects due to coordination efforts. While these risk assessments were often challenged, their ensuing budgets and schedules helped secure funding beyond the initial scopes to account for the probability of risks. This process is currently being used by FTA for large megaprojects.

Another significant achievement of the WTC projects was the fact that all work, whether at the PATH Station, any of the NYC Transit Authority's projects, or Route 9A, was completed without shutting down service to regular transportation users. We made it our goal that while building the WTC PATH Transit Hub, PATH riders would always be able to utilize the PATH Station. Regardless of the activities taking place—demolition, cleanup, excavation, construction of the Memorial, and construction of the PATH station itself—PATH ridership always had access to Lower Manhattan. Detours were constantly created. Entrances, in particular,were moved multiple times, at a cost of many millions of dollars, to ensure that riders, although inconvenienced, would always have full access to the station. Similarly, Fulton Center was phased to allow all stations within the center to continue to be utilized by commuters and tourists. Route 9A was also phased in such a way that traffic patterns would be minimally affected during construction. The most extreme case LMRO undertook was the #1 Subway line. At the location of the future WTC Cortlandt Street Station, the Port Authority had to underpin the #1 Subway-line structural box and excavate below it with intricate, top-down construction.

In conclusion, I really believe that the relationships built and the personalities on these projects made it all happen. Without our commitment to work issues through to a conclusion for everyone's benefit, the site would still be in a quagmire of bickering and indecision. Relationships were key to our success; often, calls were made for help or compromise so that progress could continue. I could only hope to be surrounded by such dedication and professionalism for the rest of my career.

Public–Private Partnerships for Rebuilding the World Trade Center

As is usual for urban disaster remediation, the rebuilding of Lower Manhattan required the interweaving of public infrastructure and private real estate, and these activities were expected to be funded from a combination of public and private sources. In this particular situation, however, the estimate of damage to the infrastructure and public domain was in excess of $21 billion, an unprecedented public cost. Further, the destruction had occurred in the heart of a leading global city with substantial disruption of private commercial places, themselves suffering physical damage estimated at above $8 billion. Therefore, the usual combination of resources for rebuilding from government and the private sector was challenged in terms of the extent of the project, the operational interconnections, and the economic interdependencies.

Of necessity, substantial modifications were made to the methods for accessing and expending public resources (see Appendix 2 for funding details). Directed by NYS, the funding was provided through either Community Development Block Grants (CDBGs) or tax-exempt Private Activity Bonds (PABs), which were renamed "Liberty Bonds." Within the framework of federal government legislation post 9-11, most specifically the Victims of Terrorism Tax Relief Act of 2001 and the Job Creation and Worker Assistance Act of 2002,[2] the criteria for utilization and methods of disbursement were modified for faster and more diverse application. This cooperative dynamic concerning the federal funding combined with legislated NYS funding, though meeting new and unfamiliar challenges, was fundamental to stimulating recovery in the Lower Manhattan areas bordering the WTC, including the revival of residential activity, job creation, new building along Fulton Street, and improved access to the southern tip of Manhattan.

Although the debate regarding aspects of public–private partnerships may continue, the rebuilding of Lower Manhattan provides an example of how such an effective collaboration is established, modified, and evaluated. Many of those who participated in the rebuilding refer to this critical feature of cooperation and inclusion in the rebuilding of Lower Manhattan.

Robin Panovka, Chairman of the real estate practice at Wachtell, Lipton, Rosen & Katz, was closely involved in the challenges of rebuilding from the very outset through the advice he provided to the firm's client, Larry Silverstein.

Rebuilding the World Trade Center: Lessons for Other Large-Scale Public–Private Projects

Robin Panovka Reflects

The rebuilding of the WTC site after September 11, 2001, now almost completed, was accomplished through the combined efforts of public and private interests, albeit not without some acrimony along the way. The remarkable "partnership" of public and private players was created out of necessity, with neither the public nor private interests able to carry out the project alone or willing to step aside. Despite the unusual circumstances, the rebuilding of the World Trade Center provides valuable lessons for other large-scale city-building projects, which almost inevitably require collaboration, or at least peaceful coexistence, by multiple public and private players.

The main players in the World Trade Center rebuilding effort were Silverstein Properties, the private developer that had leased the Twin Towers for 99 years just months before their tragic destruction on 9/11, and the Port Authority of New York and New Jersey, the public bi-state agency that owned the underlying 16 acres. But at least 30 other public and private stakeholders, politicians, and community groups, including the families of victims, also played important roles. And, not surprisingly, for a project born out of the incredible tragedy of 9/11, the many stakeholders, journalists, and commentators involved often had widely divergent views, sometimes diametrically opposed, always strongly held, and almost invariably infused with passion. Given the circumstances, the press weighed in with unusual intensity, sometimes complicating the negotiating process and often adding pressure, sometimes helpful, sometimes not.

In addition to the challenge of working with so many interests under the intense glare of the press, the site itself and the ultimate mixed-use project built were unusually complex and presented some very difficult challenges. The site was crisscrossed by subway lines, abandoned roads, and an underground PATH train line to New Jersey, which ran under the Hudson River and was surrounded by a slurry wall that created the "bathtub" that keeps the Hudson River out. And the new mixed-use project attempted to be superimposed on the pre-existing infrastructure after 9/11, consisted of a major new PATH train station connected to a new subway transit hub, the 9/11 Memorial & Museum, a performance art center, large office towers with more than 10 million square feet, and 500,000 square feet of retail space around the Calatrava-designed Oculus. Adding to the complexity, the site was encumbered by a pre-9/11 legal regime—99-year leases, multiple subleases, and a reciprocal easement and operating agreement—which applied to the

Twin Towers but was unworkable for the new project; and a regulatory structure that gave jurisdiction over the site to a number of different, sometimes conflicting, agencies.

Yet despite all this, after an initial period of jostling and a number of skirmishes, the public and private interests involved managed to work together so that in just 22 years, much of the site has now been built out.

How was this accomplished, and what lessons can be learned that might be helpful to other large-scale projects?

The answer lies in part in the unifying effect of joint tragedy and the energy generated by private entrepreneurship, but also in the remarkable power of public–private collaborations. The theory behind such collaborations is rather simple and essentially amounts to 1+1=3. Combine the strengths of the public and private sectors, align interests, create appropriate incentives, and *voilà*, a project that might otherwise have been impossible gets built faster, cheaper, and better than either sector could have done alone.

Properly structured public–private projects are a powerful tool for reimagining and re-creating cities. And as the worldwide urbanization trend continues and new infrastructure, city-building, and smart-city projects are initiated almost daily, collaborative public–private developments will inevitably play a growing and critical role.

While the theory of such collaborations is powerful, their [execution/delivery] gets bungled far too often. On the ground, things get complicated fast as the parties grapple with difficult questions: What are the real strengths each sector can (and is willing to) contribute to a particular project, and can the parties agree on them? How do you align interests and create appropriate incentives in real life, when the governance, goals, and incentive structures on both sides are so different? How do you ensure a fair outcome and not an embarrassing windfall or loss of face on either side, especially given the long-term nature of the relationship? And what do you do when something goes wrong along the way, as it inevitably will in any complex project?

This essay draws on the experience of rebuilding the World Trade Center after September 11, 2001, in an effort to glean some lessons learned that may help in future large-scale public–private projects.

Master Plan

One of the most important lessons from the rebuilding of the WTC is the critical need to reach an agreement on a Master Plan, a joint vision, as early in the process as possible. As the old Chinese adage warns, sharing a bed is not enough—to succeed, a marriage or partnership must also be based on shared dreams, or at least dreams that are not in conflict. Certainly, the different public and private players in a project might have different needs and motivations—profit versus the public interest, for example—but it is essential that both sides agree on the overarching dream, the overall vision for the project that achieves their ultimate goals. Without this, lack of alignment, infighting, and disputes are almost inevitable.

In the case of the World Trade Center, achieving a joint vision was easier said than done, and, in fact, the process took several years. The chaotic birth of the project, in the wake of the terrorist attacks of 9/11, made it unrealistic to reach a quick consensus. At the outset, the federal government took over the site under its emergency powers. There were rescue and recovery efforts ongoing and continuing concerns about security, health, and safety. There was an understandable sense that Ground Zero belonged to the public and not to any one person or institution, that it was more than just land and buildings, so that, in effect, customary private property rights were suspended for some time. There were many strongly held views about what should happen at the site, coming from many different stakeholders—property owners, tenants, politicians, families of victims, the press, and, seemingly, everyone else.

Only when the Master Plan was finally agreed to in December of 2003—under massive presure more than two years after the Twin Towers were destroyed—did the parties coalesce around a shared vision, and even then, there were divergent views on many secondary issues. The story of how the Master Plan was agreed to is instructive. By mid-2003, Larry Silverstein, the Port Authority, Governor Pataki, and most stakeholders had generally agreed on two broad principles. First, that the footprints of the Twin Towers, where so many lives had been lost, should become a memorial. And second, that it was important to reopen the Manhattan street grid (including Greenwich Street, which had run through the site before the original World Trade Center had been built in the 1960s but had been blocked by the "super-block" design of the original WTC). The reason for re-establishing the grid and the related urban planning considerations is a whole separate topic, but the important point here is that these two principles had taken hold and were strongly felt—DO NOT build on the old footprints and DO reopen the grid.

There was just one problem. The entire legal regime for the site, Silverstein's 99-year lease, and all the other ownership and leasehold stakes were completely inconsistent with these two principles. In fact, the 99-year lease required that Silverstein rebuild the Twin Towers exactly as they were, where they had been pre-9/11, exactly where everyone now agreed a memorial should be built. There was massive pressure. The Port Authority and the Governor took the position that the pre-9/11 legal arrangements had to be thrown out and that the whole situation was too complicated to sort through and renegotiate. Their position was that the project should start on a clean page, entirely controlled by government agencies, and without any private interests involved. But, Silverstein was passionate about rebuilding, and the Silverstein team believed that stepping aside would be counterproductive and lead to more delay and an inferior project.

With this backdrop, after coming back from a summit meeting with the Port Authority and representatives of Governor Pataki in October 2003, at which Silverstein had been given an ultimatum to either come up with a solution or step aside, the Silverstein team worked through the night to come up with a solution. The goal was to formulate a short and to-the-point written proposal by the following morning so that we could cut through the deadlock, show that the problem could, in fact, be solved, and start shaping people's thinking about the path forward.

What we came up with was the idea of the "immediate swap," which we presented to the Port Authority and the Governor of New York the following morning in the form of two short paragraphs and a color-coded diagram. Essentially, what we said was that Silverstein would swap his leasehold for the Twin Towers—so their footprints could become a memorial—for a leasehold on five adjacent sites surrounding the Memorial, with development rights to build the same quality and quantum of office space in the five towers as had existed in the Twin Towers before 9/11—"10 million square feet of commercially viable, Class A, above-grade rentable office space ... [each building having] an at-grade lobby." We slapped a three-colored diagram on the back that showed what was to be transferred to the Memorial in green, what would go to Silverstein for office space in blue, and what would become a train station and infrastructure in yellow. We agreed to some negotiating principles to fill in the details later, including that everyone would act reasonably and in good faith to negotiate all other issues. That was basically it. Every word was chosen carefully, but it was short and easy to understand. The key was that this was a final, binding agreement, not an aspirational letter of intent that may or may not be implemented. There was no turning back.

Once this agreement (in the form of a short letter) was signed on 12/1/2003, the swap was immediately effective, the Master Plan for the site was legally adopted and final, and rebuilding could proceed. It was a risky move, given the large number of unresolved details and issues, but it was essential to breaking through a paralyzing deadlock.

Clarity on Public Purpose

An essential element for gaining acceptance for the Master Plan and for Silverstein's central role in leading the project was articulating why everyone would be better off in this scenario. There were a number of voices early on urging the government to take over the site permanently and to exclude Silverstein from the rebuilding process. To counter this, the Silverstein team developed a powerful argument for Silverstein's central role, which we encapsulated in a mantra: "We have the *right*, we have the *responsibility*, and we have the *resources* to rebuild." It sounded simple, but each phrase articulated a carefully constructed position backed by legal, moral, political, and financial analysis and strategic thinking. Each component—the right (a reminder that there was a binding 99-year lease in place that gave Silverstein the right to rebuild), the responsibility (both contractual and moral, as the appropriate response to terrorism), and the resources (a reference to the billions of dollars of insurance proceeds to which only Silverstein was entitled and which were essential to rebuilding) —was essential to winning over Silverstein's doubters, and it ultimately succeeded.

While the circumstances were unusual, there is a risk in any public–private project that questions will arise as to whether the private developer is necessary, whether the developer is getting too good a deal at the expense of the public, and whether the public agency would be better off just going it alone. These issues should be addressed early on and should be kept in mind throughout the life of the project lest they damage important relationships required for getting the project built or even derail the project. Public support is essential and should not be taken for granted. Care should be taken not only to negotiate a fair deal but also to be prepared to articulate why it is fair when the inevitable questions are asked. Any appearance of a windfall for the private developer at the expense of the public can be fatal.

Proof of Concept

In multiphase projects that spread over many years, it may be important to demonstrate the project's feasibility and economic viability at an early stage in order to give everyone the confidence to proceed to further phases. In the case of the WTC project, the rebuilding and lease-up of 7 World Trade Center (7 WTC) served this critical role.

Rebuilding 7 WTC, which started in 2002 when nothing else was moving at the site, was essential to demonstrating that the entire WTC rebuilding effort wasn't hopelessly deadlocked, a widely held impression at the time. The collaboration on 7 WTC also created a model for resolving future issues on the larger site and demonstrated the economic viability of rebuilding. The story of how and why 7 WTC was rebuilt first is interesting and offers many lessons. 7 WTC was the last building to go down on 9/11, and fortunately, no one died there, so it was less controversial to rebuild. In fact, there was a need to rebuild 7 WTC quickly because its base included a ConEd power substation that was needed to power Lower Manhattan. ConEd, the utility that provides electricity in Manhattan, had temporary transformers spread around Lower Manhattan after 9/11, which got the job done but were clearly just a stop-gap measure. There was a real need to rebuild the transformer at the base of 7 WTC or find some other solution and quickly.

ConEd made it clear that its priority was restoring the substation at the base of the building, not rebuilding the entire office tower above it, but it was prepared to respect Silverstein's rights to rebuild the office tower as long as that didn't slow down the process or get in the way of building the substation.

While the need and direction were clear, there were some major obstacles that needed to be overcome. First, the pre-9/11 7 WTC had straddled Greenwich

Street and completely blocked it, and there was a strong feeling by many people involved, including Larry Silverstein, that Greenwich Street should be reopened and that, since the Master Plan had not yet been agreed upon, whatever was to be built at 7 WTC should not interfere with the ultimate Master Plan. This was a simple proposition, but it was on a collision course with an equally simple fact. If the new 7 WTC were to be built on only one side of Greenwich Street so as to reopen the street, the building's footprint would be too small to accommodate an office tower. The solution was to expand the footprint into another adjacent street, the NYC-owned Vesey Street, but here, the Silverstein team encountered another hiccup. While the City was willing to trade the required strip of Vesey Street for the reopened Greenwich Street, legally, it couldn't just convey the strip to Silverstein. Rather, a series of complex regulations needed to be addressed by having the required strip condemned, which was a court process that would take some time and could be challenged. There was no certainty it could be accomplished, and in any event, it would take much longer than circumstances allowed. Silverstein needed to build right away and couldn't wait for the condemnation. After weighing the risks and getting what amounted to a handshake agreement from Mayor Bloomberg that somehow the "required strip" of Vesey Street would be transferred to the Port Authority and Silverstein, Silverstein stepped up and took the extraordinary step of building a large tower on a strip of

11.1

Re-establishing Greenwich Street through the WTC site.

Source: Joe Woolhead.

land he didn't own or lease, ultimately getting most of the building up by the time the strip was acquired. It was a calculated risk that paid off. Luckily, Silverstein never had to find out what would have happened if something were to have gone wrong.[1]

The rapid rebuilding of 7 WTC faced another complex challenge. The rebuilding of 7 WTC included the ConEd power substation mentioned earlier, in the first 77 feet, with the office tower directly above it. Construction was to be on a super-fast track before the plans for the building were completed, and it was not possible to allocate costs between ConEd and Silverstein in advance of building. There was no way to determine how much of the foundation, elevator banks, façade, and other elements were attributable to the office versus the substation until full plans and cost estimates were completed, and there was no time to wait for the final plans. So, given the urgent need to rebuild, Silverstein elected to proceed based on weekly agreements rather than wait until a comprehensive approach was possible. Each week during construction, Silverstein and ConEd, with the help of the general contractor Tishman Construction, entered into a simple agreement allocating costs for the prior week and agreeing on what was to happen the following week. No one knew what would happen if Silverstein and ConEd couldn't reach agreement the following week, and luckily, we never found out.[3]

11.2

7 WTC with ConEd substation on lower floors.

Source: SOM/dbox/Silverstein Properties.

Stick to Your Guns, or Pivot?

Unexpected issues and obstacles are sure to crop up in any multiparty, public–private project. One of the key skill sets in dealing with such surprises is the judgment to decide when and how to be flexible and pivot. Insistence on sticking to the original plan is often helpful to avoid mission creep. But, on the other hand, facts on the ground change over time, and excessive rigidity can also be damaging.

Moving the site of 1 WTC is a case in point. Two years after the Master Plan had been signed by all parties, including the City of New York, the New York Police Department sent a letter to the Port Authority noting that the planned 1 WTC was too close to the West Side Highway and, as such, unsafe. A greater distance was necessary to protect the building from truck bombs and the like, which could easily be driven up on the West Side Highway. The parties were outraged. The location being complained about by NYPD was the exact location shown on the Master Plan that the city had approved. And at least $10 million had been spent on developing plans for a building that now would have to be moved away from West Street and, therefore, redesigned. Why did NYPD just wake up? Did the Port Authority know of NYPD's concerns, as some claimed, but had decided to ignore them? Was the NYPD, in fact, right about the safety concerns, and if so, why did the Port Authority's safety experts not raise the same issues? Should the NYPD be challenged? These kinds of questions swirled around for a couple of weeks while the project stalled. And then, after some venting, it became clear to the Silverstein team that none of these questions really mattered. The smartest course was to pivot, quickly redesign the building to satisfy NYPD's concerns and get on with the project. While we could have insisted on sticking with the original plan, it became clear that building a tower that NYPD publicly declared was unsafe would be foolish. Who would lend against such a building? Who would insure it? Who would lease space in it? Silverstein pivoted quickly and moved on.

Keeping It Simple, One Step at a Time

Complex projects often have too many moving pieces to tie everything down at the outset and instead require a stepped approach. At the outset of the WTC project, no one knew what was going to be built (there were multiple decision-makers with widely divergent visions), how much money was available to build (the insurance recovery process was just getting started), and who would control the rebuilding (the Port Authority or Silverstein). There were just too many unknowns for any kind of comprehensive agreement. Instead, the parties entered into a series of binding term sheets and letters of intent, agreeing on general principles and the details that had been worked out and leaving the rest to future agreements. As details became known and as issues got negotiated, the agreements grew from short letters and term sheets into full-blown development agreements and leases that spanned thousands of pages. If the parties had insisted on full documentation before starting, the project would never have gotten off the ground.

Negotiating Team

Essential to the success of the WTC negotiations was the formation of a small, collaborative Silverstein–Port Authority negotiating team that was able to maintain a sense of balance and humor and not overreact to the PR crisis *du jour*. Significant PPPs inevitably have political, press, and public relations components, which can become highly charged. These components can (and should) be helpful if managed properly, but they can also be distracting and counterproductive. Over the course of many months and years, the WTC negotiating team developed mutual trust and respect, was sensitive to the political and PR pressures on each of its members, but kept its eye on the ball, getting the larger deal done.

No Thumb on the Scale and Other Protocols

In the WTC negotiations, the negotiating team developed a set of unwritten, informal protocols and rules. One was the "no thumb on the scale rule," which required leaving future issues unresolved and undisturbed. When anyone tried to add language to an agreement that created a bias in favor of a particular approach to solving an unresolved issue, others would remind him or her that they were placing a thumb on the scale, which was inconsistent with the rule. The idea was that at any given stage in the process, we would resolve only what needed to be resolved and could be resolved at that time, with other issues deferred by articulating even-handed, broad principles for future negotiation. It was not possible, for example, to agree on all of the terms of redevelopment when we signed the 12/1/2003 "immediate swap" letter—there were too many moving pieces, too many unknown and unknowable facts—so we had to proceed in incremental steps. It was understood that after each incremental step, no one would try indirectly to influence the outcome of future negotiations, but rather that we would table the issue until it was ripe for resolution. Any attempt to slant drafting or facts on the ground in a way that would create future advantage, or any collateral agreement that would prematurely change the status quo, was an impermissible "thumb on the scale," entitling the other parties to cry foul.

It Took a Lot of Mutual Respect and Trust to Make This Work

Efficient Dispute Resolution

Litigation is a killer. If the partners in a project end up in court, the resulting delay and animosity may well spell the death knell for a project. Litigation can take many years if appeals are pursued, and few projects can absorb that kind of body blow. The solution crafted into the WTC documents was super-fast-track arbitration after a quick period of forced negotiation by the parties' respective CEOs. In essence, the documents provided that if a dispute arose, the CEOs of the warring parties would have to meet in person over a period of a few days, and if they failed to resolve the issue, accelerated arbitration would follow with a decision required within a short period. The parties ended up using these provisions to resolve some thorny situations. The process was quick, relatively amicable, and enabled the project to proceed without undue delay.

Know Your Counterparty

Public and private players think differently and are populated by different kinds of executives. Sophisticated parties take this into account, try to understand each other's perspectives and goals, and, most important, try to craft solutions that are wins for both sides. A truly successful project satisfies the private developer's economic expectations and is profitable, but at the same time, satisfies the governmental counterparty's political and institutional goals. At every level, it is critically important to understand the motivations, incentives, time parameters, and risk tolerance of the other players and to recognize that they will often be different. A government official will often care much less about costs and profits than a developer and far more about safety, security, and political elections. There's nothing wrong with having different profiles, and, in fact, it is the combination of the two that creates a stronger team. The results and the WTC bear this out.

PR Matters

The press can make or break a high-profile project that requires government support. Attacks or unfriendly scrutiny from the press can create tremendous pressure, particularly on government officials seeking re-election or accountable to people who are. A well-defined, hopefully, joint PR strategy is essential and should be viewed as a core element of any large project where public support is required. In the case of the WTC, the press was sometimes used as a weapon to exert pressure and [at] other times as a tool to help build support for the project. Either way, this is an essential skill set in any high-profile project.

*****

All in all, the WTC project demonstrated the importance of arrangements that aligned the parties' interests and created long-term incentives to perform—not so simple when you're dealing with as many diverse stakeholders and motivations as existed at the World Trade Center; the importance of building trust and negotiating protocols among the negotiators; getting into the heads of counterparties in order to understand exactly what they are trying to achieve and to help them get there; understanding that in complex multiparty terrain like Ground Zero it is impossible to please everyone, to find the perfect solution or to draft the perfect agreement, so it is often wise to seek Pareto optimality and avoid getting bogged down trying to achieve perfection; trying to resolve disputes and deadlock relatively quickly before they spin out of control or spill into the press by changing the dynamics or parties at the table, seeking third-party intervention, forming alliances or through PR; getting out ahead of complexity and shaping people's thinking with simple, bold proposals like the "immediate swap"; respecting "off the record" discussion; and avoiding *ad hominem* attacks if at all possible.

Much of negotiation is art, not science, like judging when to stick to your guns and when to compromise, when to draw lines in the sand, and when to roll with the punches, and the applicability of any rule will depend on the specific facts and circumstances. 7 WTC might never have been built if Silverstein had not been willing to give up Greenwich Street and build on a week-to-week basis on a city street he didn't own, and the Master Plan might never have been adopted if Silverstein had not been willing to agree to the "immediate swap" in which he gave up well-defined, multibillion-dollar leases in exchange for a few (admittedly carefully crafted) paragraphs with a basic outline and principles for future negotiation.

*****

In the end, the results at the World Trade Center hopefully speak for themselves and prove that 1+1 can really equal 3. The centerpiece is a dignified memorial, museum, and public space that hopefully will serve generations to come, and the perimeter includes modern office towers and stores that create jobs and support the commercial life of the city. The site has once again become a thriving part of the city, arguably better designed and more accessible than before. The compromises and diverse interests represented have created a uniquely rich environment.

Notes

1 Daniel Creighton, PE, is Vice President of MTA Capital Construction and former Director of the FTA Lower Manhattan Recovery Office (LMRO).

2 H.R.2884 — 107th Congress (2001-2002) - Victims of Terrorism Tax Relief Act of 2001; H.R.3090 — 107th Congress (2001-2002) - Job Creation and Worker Assistance Act of 2002.

3 Refer to Alex Garvin's explanation of this situation on pages 38–40.

Chapter 12: Economic Progress in Lower Manhattan, to 2019 and to 2022

Impasses

All aspects of the Lower Manhattan economy flourished during the 2010s under the umbrella of phenomenally strong growth in NYC's economy, where the pace of job expansion exceeded that of the U.S. economy.

By 2019, the Lower Manhattan economy had finally reached its pre-9/11 levels. Private-sector employment exceeded 250,000, and there were over 56,000 public-sector jobs. The population had surpassed 64,000 by the time the decennial census was taken in April 2020.

Whether in office leasing and development, hotel and residential development, the arrival of new Michelin-starred restaurants and other retail, and the creation of new parks and initiation of critical projects for climate resiliency, by year-end 2019, Lower Manhattan was a thriving 24-hour community and clearly had become the third or fourth largest CBD in the United States.

The Office Sector

Office leasing in 2019 was at the highest volume since 2010, with 7.3 million square feet leased, the largest relocation leases signed by Uber and Justworks, a human resources technology company. As of 2019, financial firms were still critical to the Downtown economy, as the largest private-sector companies were Goldman Sachs, Deutsche Bank, Morgan Stanley, American Express, Bank of America, and American International Group (AIG). The Alliance for Downtown New York noted in their 2019 report, "As other Manhattan submarkets have become increasingly expensive, Lower Manhattan continues to offer tenants quality, modern space at competitive prices along with unparalleled transit access."[1]

The Downtown office inventory in 2019 was 21 percent of the total Manhattan inventory, which is primarily located in Midtown. Further to the Downtown Alliance 2019 report,[2] this much-improved inventory in Downtown is still below the level of 107.7 million square feet measured in the summer of 2001, which was then approximately 25 percent of the Manhattan inventory, but it should be recognized that over the past two decades, many older office buildings in Downtown have been converted to residential use. Continuing with the opportunity for conversion under the 421-g legislation, a number of more historic, smaller office buildings have sold for around $400 per square foot.

DOI: 10.1201/9781003476139-12

12.1
1 WTC opened in November 2014.
Source: Joe Woolhead, courtesy of Silverstein Properties.

A dedicated conversion developer, Metro Loft, has converted over 3.5 million square feet across 11 buildings.

Rental Rate Trends

While financial firms still were dominant, one of the characteristics of leasing and relocation of firms during the 2010s was the arrival of firms in the newly designated "TAMI" industries (Technology, Advertising, Media, and Information). This was noted at a seminar the co-authors hosted on April 10, 2017. In reflecting on the evolution of office construction at the WTC site, Janno Lieber, then President of World Trade Center Properties, Silverstein Properties, noted that by 2016, TAMI represented 48 percent of the tenants at the WTC and that 75 percent of the recovery in office tenants "is not coming from the financial sector."[3] This trend continued throughout the rest of the decade, and by 2019, such firms as WeWork, GroupM, Spotify, Knotel, Holltzbrinck Publishing, Omnicom, and Uber had arrived to join the early newcomer, Condé Nast in this TAMI sector.

Flexible Space

As the area has accommodated these new types of tenants, the format of workspaces has changed also. Although WeWork provides a high-profile example, the spatial and operational conditions for the technology and creative industries have evolved to be flexible and shared. Keen to attract these new tenants, office landlords either offered flexible space alternatives or, more often, gave master leases to known operators such as WeWork and Knotel, and also engaged the public place and amenities operator, Convene.

By 2019, WeWork had 14 locations in Downtown, comprising 1.3 million square feet, and Knotel had 10 locations, totaling 0.5 million square feet, in Lower Manhattan, while many new buildings provided the new configuration to tenants themselves. Convene has taken up space in Brookfield Place and other buildings, with almost 300,000 square feet in the area.

Despite the strong leasing of offices and the arrival of tenants new to the Downtown market, in its year-end report for 2019, the Downtown Alliance noted the persistence of large blocks of space still available along Water Street, at the WTC, and at Brookfield Place.

Tourists to Downtown: An Important Boost to the Economy

The year 2019 was also a strong year for tourism, with some 14 million visitors coming to Downtown. The influx of business and tourist visitors prompted a surge in hotel activity, with almost 8,000 rooms in 37 hotels available by year-end 2019, with another 2,000 rooms in 15 hotels in the pipeline.

The development of the residential sector has been truly remarkable during the decade, ending the year with an estimated 64,000 residents living in 33,247 units in 337 mixed-use and residential buildings.

As a result of the residential, employment, and visitor growth, the area's restaurant industry has not only grown in size but also in notable quality, as two of its recent arrivals have earned Michelin stars. In addition to restaurants, new retail stores opened throughout the neighborhood, including food markets, bookstores, and fitness centers.

12.2

Tourists visiting the Memorial.

Source: Joe Woolhead.

New infrastructure investments include the ongoing plan to extend PATH capacity at the WTC Station to nine from eight cars and improved capacity to Newark Airport as the Port Authority completes its planned 2.4-mile extension to the airport from Newark Penn Station (and, thus, from the WTC). In addition to ongoing street reconstruction, the Robert R. Douglass Pedestrian Bridge over West Thames Street to Battery Park City was completed in the fall of 2019. Planning or construction is also underway for several new parks, plazas, and greenways throughout the Downtown area.

Critically important for the future of all of Downtown are the new plans for investment to bolster resilience in the face of rising sea levels and more extreme storms. The lessons from the damage caused by Superstorm Sandy have been heeded by city officials. Mayor de Blasio announced in March 2019 a plan to extend the Manhattan shoreline on the East River by 500 feet with "high points at or above 20 feet." Future plans will extend the shoreline for the Financial District.[4]

Rebuilding the Residential Community

An important component of the rebuilding effort has been the residential development. Though small for much of the 20th century, while the area served as a global finance hub, in the later decades, housing was encouraged.

Overriding Objective for Lower Manhattan Land-Use

As a result of the many studies and following numerous discussions and meetings, the LMDC importantly clarified the intent to use the WTC site to knit together the community of residential plus commercial tenants.

At the same time, the legislative mandate of the land use was to continue as a revenue source for the Port Authority, and commercial rebuilding of 10 million square feet was necessary.

And in contrast to when the original WTC had been erected, the contemporary trend for businesses is to locate in vibrant urban neighborhoods. Therefore, to bring businesses back, Lower Manhattan needed a 24-hour community that would include some new residential and recreational options with the new office buildings.

Resolution for Residential Rebuilding

The LMDC's objectives for the residential rebuilding activity were:

1. Retain and rebuild existing residential units.
2. Continue the conversion of obsolete office buildings using the 421-g program.
3. Encourage new building—this would require some incentive programs to stabilize rental levels for development feasibility and to support funding of this risky undertaking in the context of weak capital markets.

With the rebuilding, the residential building stock has increased significantly, and the amenities needed by that influx of residents are being provided.

The scale of the neighborhood, the retention of some potent historical structures complemented with contemporary residential structures suitable for urban families, and the potential of existing retail markets such as those along Fulton Street and South Street Seaport provide an attractive urban fabric for the creation of a vibrant inner-city residential neighborhood.

Achieving a population of 64,000 today represents a 156 percent increase since 2001.[5] Today, most inhabitants are young to middle-aged adults: 50 percent are aged 25–44, and 18 percent are aged 45–64. With 14 percent of the population aged 0–17, the area has been popular with small families.

Challenges

However, adding this residential population to the area presented challenges:

- *Social services*: There was a need for the substantial expansion of social infrastructure, such as medical centers and schools, as well as open space.
- *Domestic goods and services*: Increased retail space was needed, including supermarkets and other domestic suppliers.
- *Urban equality*: The economic diversity of the residents was skewed toward the very high end since the housing rental rates were unaffordable to many New Yorkers, and to provide equal access to housing in the area, subsidized units were necessary. However, an estimate of the required subsidy for a standard family unit was $400,000, given the cost of the local land component, and, therefore, if 20 percent of the new units (900 units) were to be delivered as affordable, the total capital subsidy would be approximately $360 million. Further, the question is "affordable for whom?" Is that 40 percent of Average Median Income (AMI)—the measure of household income—or 120 percent of that AMI; that is, 40 percent of AMI being severely low-income, in contrast to 120 percent of AMI being better-than-average income for NYC?

Outcomes

- Residential development exceeded expectations.
- The residential community's presence encouraged the business community to stay.
- But only the higher income levels of residents are located here.

By 2020

Going into 2020, Lower Manhattan had 33,247 residential units in 337 mixed-use and residential buildings.[6] Continued growth was anticipated, with 2,800 units in 17 buildings under construction or planned for development.[7] Of this pipeline, approximately one-third were rental units, and two-thirds were condominiums for sale.

Rental rates continued to increase significantly during the decade, though this is attributed, in part, to the expanding inventory of new, amenity-rich luxury units.

Despite its changing trends in rental rates and sales prices, the rebuilding of the residential sector in Lower Manhattan has proven to be spectacularly successful.

Impact of the Pandemic: 2020–2021

After the remarkable growth and transformation of Lower Manhattan, first in the years following the catastrophic events of 9/11 and then after the painfully disruptive impact

12.3
30 Park Place Hotel and residences.
Source: Joe Woolhead.

of the financial crisis and Great Recession of 2008, came the worldwide onslaught of the Coronavirus known as Covid-19, an onslaught on lives, on economies, on everyday patterns of work, shopping, dining out, or taking a vacation. The economy and everyday living in Lower Manhattan, as in the rest of NYC and the rest of the world, came to a standstill. Office leasing plunged from over 7 million square feet in 2019 to just over 2 million square feet in 2020 and barely eked out 2.8 million square feet in 2021. Office vacancy rates rose to 21.6 percent, the highest in Manhattan. Over 100 retail stores and restaurants closed, and job levels fell to 230,000 private-sector employees from a high of over 250,000 in 2019.

Despite the citywide gloom of intense lockdown and resurgent Covid variants, there were some hopeful signs by late 2021 and early 2022. The Wall Street Hotel at 88 Wall Street, with its high-end restaurant, *La Merchande*, opened in late 2021, as did the Saga restaurant at 70 Pine Street. Jean-Georges opened at the Tin Building, and several food stalls opened at Hudson Eats at Brookfield Place. New retail openings were also expected at Brookfield Place and Westfield (the Oculus). Tourists were returning by the summer of 2022, some eight hotels with 1,500 rooms were in the pipeline, and over 4,000 residential units were in construction or development. The St. Nicholas Greek Orthodox Church opened in late 2022, and the Perelman Performing Arts Center opened in September 2023.

12.4

The Oculus deserted.

Source: Joe Woolhead.

In Downtown, restaurants are again open, retail stores have increased foot traffic, and the use of the subway, bus, and PATH systems has rebounded, though at mid-2023 still not yet to pre-pandemic levels. With the return of tourist and business visitors, Lower Manhattan is again contributing importantly to the NYC economy.

It is too early to surmise how seriously this pandemic will affect the office market of NYC in general, and in Downtown in particular, whether in the near or intermediate term. As many office buildings have begun to reopen, some office workers appear reluctant to return to their former offices despite careful protocols of spacing, elevator occupancy, and regular cleaning. Fear of commuting on crowded trains or buses may also keep workers at their new-found WFH (work from home) setups. In a survey[8] of 140 major Manhattan office employers between August 23 and September 15, 2023, 58 percent of workers were in the office post-Labor Day (September 4). The stated expectation of employers was that this trend would not change significantly.

Notes

1 The Alliance for Downtown New York, *Lower Manhattan Real Estate Year in Review 2019* (2020), p. 2.

2 The Alliance for Downtown New York, *Year in Review 2019.*

3 Janno Lieber, panelist at T+15 symposium session on "Envisioning the Office Market Downtown, Post 9/11," April 10, 2017.
4 In August of 2022, FEMA provided NYC's Economic Development Corporation with $50 million of funding for the Financial District and Seaport Climate Resilience Master Plan.
5 The Alliance for Downtown New York, *Year in Review 2019*.
6 The Alliance for Downtown New York, *Year in Review 2019*.
7 The Alliance for Downtown New York, *Year in Review 2019*.
8 Partnership for New York City, *Return to Office Survey*, September 2023.

Chapter 13: Finalizing the Last Major Projects of the World Trade Center Redevelopment

Cortlandt Street Station for #1 Subway Line

Major engineering and construction work was required to rebuild the Cortlandt Street Station on the #1 Subway line.

Daniel Creighton Reflects

The reconstruction of the WTC Cortlandt Street Station required intricate planning and perseverance and could not begin until surrounding WTC infrastructure was completed. In the aftermath of 9/11, a top priority was to restore the #1 Subway line service to South Ferry Station while plans to rebuild the overall WTC site were still taking shape. Construction on the station itself could not even be fully designed until this work was completed.

Given that the station sits in the middle of the WTC site, each element of the redevelopment directly impacted the planning for the new station's entrances and overall design. Its connection to the area's new urban retail space makes its location all the more integral to the new urban fabric that was created during the entire WTC redevelopment. The location also meant working with numerous state, city, and private stakeholders to accommodate construction schedules as well as maintain overall design standards.

At the end of the process, 1,000 linear feet of subway tunnel were elevated on a structural steel bridge structure, which cost several hundred million dollars so that #1 Subway-line commuters to and from South Ferry Terminal would continue to have uninterrupted service. This shows the extent that the many authorities and state agencies would go to in order to minimize construction impacts on the public.

The new WTC Cortlandt Street Station, completed in September 2018, brings #1 Subway line service three blocks closer, making it significantly more convenient for thousands of WTC complex employees and visitors. This was accomplished by fully integrating the reconstruction into the site planning of the WTC project.

DOI: 10.1201/9781003476139-13

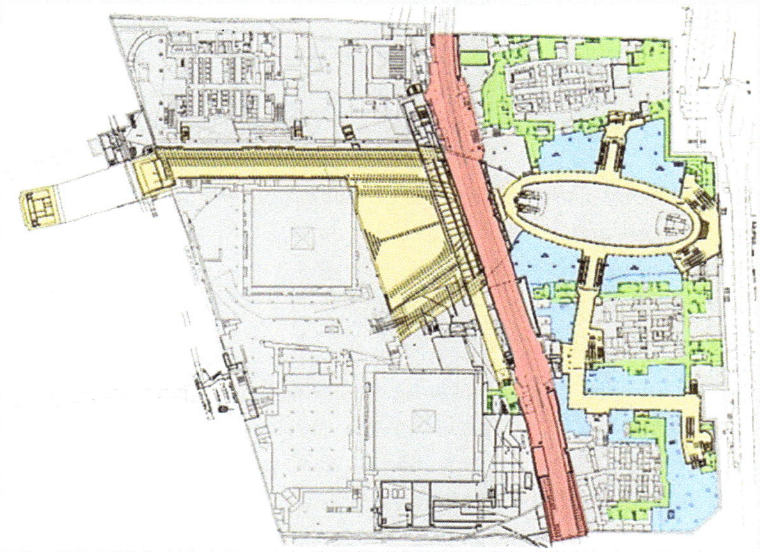

13.1

The #1 Subway line through the WTC site.

Source: Metropolitan Transportation Authority.

13.2

New Cortlandt Street Station for #1 Subway line.

Source: Joe Woolhead.

The station features full ADA accessibility [for people with disabilities], is air-tempered to keep it comfortable year-round, and it is equipped with state-of-the-art security, communications, and life-safety systems. The aesthetically modern station links directly to the World Trade Center Transit Hub (the Oculus) with connections to 10 NYC subway lines, PATH service, and direct access to the WTC's office towers and retail facilities.

There was such pride and a sense of accomplishment in the opening of the new Cortlandt Street Station for the #1 Subway line, and Sandra Bloodworth, head of the MTA Arts and Design program, arranged for a celebration combining the revival of the infrastructure with the blossoming of arts and culture in Lower Manhattan. She commissioned a site-specific art installation, *CHORUS*, by Ann Hamilton for inside the station.

Later in the redevelopment progress, at the Port Authority, Carla Bonacci[1] led the oversight of project managers and construction inspectors for St. Nicholas Greek Orthodox Church (completed and opened in December 2022) and the Perelman Performing Arts Center (opened in September 2023).

St. Nicholas Greek Orthodox Church and National Shrine

St. Nicholas Greek Orthodox Church was completely destroyed in the collapse of World Trade Center Tower Two during the September 11, 2001, terrorist attacks. In the weeks and months that followed, there were funerals and memorials for the many Greek Orthodox Christians who died that fateful day, as well as interfaith and ecumenical events at city, state, and national levels.

The design for the rebuilt church at the southern boundary of the WTC site is by Santiago Calatrava. The structure of the St. Nicholas Church itself cost $85 million and features white marble imported from the same quarry that provided stone for the Parthenon in Athens.

Rebuilding had been stalled due to planning conflicts and funding challenges until, in early 2020, the NYS Governor announced the resumption of construction. The church was completed and officially opened on December 24, 2022.

13.3

The rebuilt St. Nicholas Church.

Source: Silverstein Properties.

Rebuilding Culture

In addition to the important 9/11 Memorial & Museum as a vibrant urban hub, the WTC site includes a venue for cultural performances and events. Of the $21 billion in federal funds provided for rebuilding Lower Manhattan, $200 million was earmarked to support cultural institutions, and this provided seed money for numerous institutions, small and large, in the neighborhood as well as an important, symbolic venue on the WTC site itself. This was envisioned in Libeskind's site plan as a performing arts center.

The Perelman Performing Arts Center

Construction has just been completed on the striking cube-shaped venue for performing arts at the northeast corner of the WTC site, with its opening date during the 2023–2024 cultural season. Substantial funding for the Perelman Performing Arts Center (PAC NYC) was provided by the NYC philanthropist Ronald O. Perelman, as well as by former Mayor Michael Bloomberg.

The PAC NYC offers three theaters of varying sizes, within which a wide range of seating configurations and performance formats are possible. The internationally acclaimed architectural firm for the project is REX, founded and led by Joshua Ramus and based in New York.

Partnership for Programmatic Details and Evolution of the Design

The determination of programmatic details and the nature of the design for such a progressive venue on the symbolically important WTC site, together with the complexity

13.4
The Perelman Performing Arts Center.
Source: Joe Woolhead, courtesy of Silverstein Properties.

of its philanthropic funding, was another situation in the rebuilding of Lower Manhattan where the need for the engagement and constructive partnership of various stakeholders was required. This was addressed by the following consultative structure:

1. Maggie Boepple, President, led the creation of the detailed programmatic requirements. She also played an important role in creating and fostering the design development with REX.
2. An advisory group of 10–15 people was formed to include representatives from the theater, performance art, directors, artists, administrators, scene and costume designers, and others. This collection of diverse professionals was led by the Consulting Artistic Director David Lan, the Artistic Director of London's Young Vic Theatre.
3. There were numerous other entities from which various major approvals were required: the Port Authority, LMDC, NYS, NYC, NYPD, and the Federal Counter-Terrorism Task Force.

Building Form

The building is situated 70 feet from the North Pool. The simple cube—160 feet x 160 feet x 115 feet—belies the complexity of the subterranean infrastructure with a depth of 21 feet, providing access to the underground areas across the WTC site, including security checkpoints. Frank Gehry was the first architect for the project and designed the building to fully span the site since below is the vehicular access to the WTC underground area.

Later, taking on the project, REX twisted the building form from its original footprint to achieve a solution that resolved the differing pressures on the project—its programmatic needs, the foundation constraints, its interplay with other forms of the WTC and security requirements. This placed the main stairway facing toward the Memorial. Ascending or descending the 21-foot staircase to the center affords a moment of repose and visual interaction with the Memorial. From this angle, the purity of the form of the Memorial is fully appreciated, and a counterpoint is provided by the rising cube with the sunken void of the North Pool.

Joshua Ramus describes to the authors his intent for this aesthetic and symbolic dynamic: "A respectful dialogue occurs. The exterior is sober and respectful in the simplicity and stability of its form, but it also hints at the flexibility of its interior and the dynamic nature of its use."

Although apparently lightweight, the superstructure is a cable net of narrow steel beams, and all hung from above. Any potentially damaging impact on the façade from outside or inside—in the event of a future terrorist attack—is absorbed by the hanging screen of the cable net and panels. At night, the form dissolves as the interior light glows through the semitransparent façade so that the activity within is conveyed without being voyeuristic.

Leslie Koch, who served as the President of the PAC NYC from 2019 until 2022 and since then as its President of Construction, commented:

> The Perelman Center completes the vision for the rebuilding of the World Trade Center site, which began with the Master Plan in 2003. ... We will celebrate all

13.5
WTC plan showing the twisted siting of the PAC.
Source: REX Studio.

13.6
Rendering of the PAC.
Source: REX Studio.

13.7
The PAC overlooking the Memorial Pool.
Source: REX Studio.

that is New York in the fields of dance, music, theatre, and everything that happens live.[2]

Plans to Finalize the Office Towers at the New World Trade Center

2 WTC and 5 WTC

At year-end 2023, the plans by the architectural firm Foster + Partners for 2 WTC were on hold while Silverstein sought the commitment of an anchor tenant.

Similarly, plans for the redevelopment of 5 WTC, the former Bankers Trust/Deutsche Bank Building, are on hold pending the completion of financing.

The proposed residential tower, located adjacent to the original WTC site, will be developed as a partnership between the public sector—comprising New York State, the Port Authority, and the Battery Park City Authority; and the private sector—including

13.8
Rendering of 2 WTC (at center).
Source: Foster & Partners, courtesy of Silverstein Properties.

13.9
Demolition of the former Bankers Trust/Deutsche Bank Building.
Source: Joe Woolhead, courtesy of Silverstein Properties.

13.10
Proposed 5 WTC.
Source: Kohn Pedersen Fox.

Brookfield Properties, Silverstein Properties, Omni New York, and Dabar Development Partners. It will feature market-rate and affordable rental housing, condominiums, a hotel, and ground-level retail.

Notes

1 Carla Bonacci AIA, PP, had more than 30 years of continuous involvement as a planner, architect, and builder at the WTC. From 2002–2022, she was Assistant Director of WTC Construction at the Port Authority.
2 Leslie Koch blog in Downtown Alliance newsletter, January 18, 2021.

Chapter 14: Lessons Learned in Rebuilding Lower Manhattan

The tragic event of September 11, 2001, that necessitated the rebuilding of Lower Manhattan, although unique in its geopolitical origins and ramifications, does share much with the social and economic crises and environmental disasters that increasingly devastate urban centers around the world. The consequences of these disasters are a tight complexity of physical and financial issues that challenge the capability of communities and cities to regroup and rebuild without great pain and dislocation. The stakeholders who are affected, the responsible administrations, and the agents of urban production—businesses, developers, and social services—are not easily brought together to achieve revitalization, and the resulting conflicts often become the major impediment to an optimal outcome for all. This was immediately and publicly apparent in the rebuilding of Lower Manhattan, though the negotiations, resolutions, partnerships, and collaborations over the past 20+ years, which have produced the vibrant WTC in 2023, provide many useful lessons.

Urban Collaboration

Contemporary urban development scholarship addressing the revitalization of city centers provides substantial guidance for planning authorities and legislative leaders. It was immediately apparent that the future rebuilt environment of Lower Manhattan necessitated the inclusion of developers and commercial properties along with governmental entities. Not only would the rebuilt area include privately built and owned buildings, but the physical complexity of the infrastructure and its substantial cost required a more intensive partnership of public and private interests than would be usual. As described, this collaboration has been the overarching theme of the revitalization of Lower Manhattan.

Multifaceted Evaluations

Not only was the inclusion of various local entities critical, but it was also necessary to achieve an understanding of how the damaged area was interconnected to the various municipalities and the wider metropolitan region. The key areas were:

DOI: 10.1201/9781003476139-14

14.1
WTC with Tribute in Light 2017.
Source: Joe Woolhead, courtesy of Silverstein Properties.

14.2
Memorial ceremony, September 11, 2023.
Source: Joe Woolhead, courtesy of Silverstein Properties.

- the contribution of economic activity, including the need to assure recovery in the financial services sector;
- rebuilding transport connections from Lower Manhattan to other parts of the city and suburbs in Northern New Jersey;
- providing the supply of residential buildings needed in the metropolitan area, and especially to provide more residential options for local workers and achieve a 24/7 urban vitality in Downtown Manhattan;

- providing support to the victims' families, which was highly charged emotionally, with solutions complex and personal;
- building a memorial to the significant loss of life.

The public, not just locally but nationally and internationally as well, was affected by both the physical devastation and the security implications of the terrorist attack. The substantial public funds involved—$21 billion in federal funds—were unprecedented, and the distribution would require navigation through the agencies of three levels of government: federal, state, and local.

The environmental urbanists, Svendsen and Campbell noted:

> This suggests the need for disaster planning and recovery models to be flexible and fluid as they respond to the needs of a community—observing the ways that people reclaim their public landscape through memorial-making. The power of these spaces resides in their civic-led and decentralized decision-making structure; attempts to support or formalize them as part of memory work must keep this foremost in mind. It is important to note that some of these efforts are fleeting, while others become longer-term building blocks for community recovery.[1]

Tim Gilchrist Highlights the Major Challenges of the Rebuilding Process

Sorting through the many battles and negotiations over the early planning issues:

LMDC was an important early player as a creation of the state and city and served as a host agency for much of the planning based upon their role as lead environmental agency for the site. But none of the agencies with construction responsibilities was willing to give up authority over their own projects to a non-construction agency.

Personal relationships needed to be established, with many players having never worked with each other. Because of our long-term involvement in West Street and recent construction in the area, we often had past relationships with major players, which helped in the process.

The public involvement process, while essential, was time-consuming and staff-intensive. The views of victims' families, local residents, and the general public needed to be taken into account and were often in conflict.

Finding agreement on the location of the main structures on the 16-acre site: the location of 1 WTC, of the Memorials in the footprints; the location of the Museum; of the towers 4 WTC, 3 WTC, and 2 WTC; the PATH Station/Oculus; the protection of the slurry wall. I was involved in other battles—the final location of the Freedom Tower or 1 WTC, the location of the Performing Arts Center, where to rebuild St. Nicholas Church, which was destroyed during the attack. Its current placement on top of Liberty Park on the Church Street side appears to be a great compromise.

The difficulty in aligning goals of the private partners—each one thought they were the most important, Brookfield, Silverstein, Goldman Sachs, and others. Their goals and wants needed to be coordinated with the public-sector priorities.

The enormity of the construction required an enormous amount of coordination, and no agency wanted to compromise on its schedule—each was ambitious, and in many cases, it was unrealistic. The changing security requirements required the redesign and rethought of the public spaces and how the traffic moved into and around the site.

The political leadership wanted everything done as soon as possible, but the work required detailed sequencing, and the needed coordination did not always easily happen. Work on the site, the Memorial, and Route 9A were interconnected. But we achieved success when we could finally get the major agencies in one room, talking to each other...

Administrative Collaboration and Integration

Scott Rechler's actions as Vice Chair of the Port Authority fell into two categories:

1. *Definition and Integration of Responsibilities and Actions*
   While the definition of responsibilities is an important first step in managing a multi-actor project, sometimes the actual process of role definition, by seemingly negating the need for coordination, has the unintended consequence of fragmenting the plans of action. As responsibilities are usually articulated by specified objectives, it is necessary that their definition not only describe their own interests but also establish key points of connection between the allocated objectives/responsibilities. It is through the acknowledgement of shared outcomes that various entities can be coerced into coordinating their activities.
2. *Align Responsibilities and Leadership with Capacity for Action*
   As the responsibilities of the various parties involved in the comprehensive effort were being defined, an explicit assessment of each with respect to their unique capacity for action was required. In the enthusiasm to make progress, some entities had sought to accomplish tasks that were beyond their resources, capability, or authority. Other entities that had certain resources sometimes refrained from contributing where they were capable due to political concerns or conflicts among leading figures.

   However, all skills and resources, at many scales, were required, and Rechler oversaw a comprehensive definition of needs and certain reallocations of resources. Additionally, he ensured that duplicative processes were eliminated and conflicting authorities were converted to integrated decision-making.

In the re-energized rebuilding of Lower Manhattan, these advancements in procedural management were particularly evident in coalescing objectives, allocating responsibilities, and assigning leadership, as well as demonstrating excellence in the design of the urban fabric.

Daniel Libeskind Summarizes the Key Design Imperatives in Rebuilding

The process must be participatory.

The product must be affordable and suitable for a mixed and diverse population.

There should be architectural diversity and innovation.

The aesthetic genre is that of collage: "Everyone is an artist creating a city in space and time."

The public will be more involved in city issues, in planning issues, like what occurred at the Javits Center during our planning process. Technology will make that more possible, as well as the actual density of New York.

Also, there must be an emphasis on affordable housing—our big cities cannot just be for one class or one type of people.

Planning for the Future

In considering the potential of Lower Manhattan beyond just its restoration, there arose some new objectives that became important drivers of the rebuilding effort. These were:

- Manhattan's CBD had become centered at its Midtown. Therefore, the connections—by mass transport and private vehicles—from there to Lower Manhattan needed upgrading.
- Brooklyn's metro area and technology hub were growing in commercial activity; therefore, Lower Manhattan required better connections to those new activities.
- The 9 to 5 office district that had characterized Lower Manhattan was outdated, and it was necessary to create a vibrant mixed-use urban community with more residential options and neighborhood amenities.
- As NYC's cultural scene expanded, Lower Manhattan could house more arts organizations through the reuse of obsolete building stock to incorporate new facilities at the WTC.
- Although served by a diversity of mass transport, Lower Manhattan's stations were often obsolete, undersized, and fragmented, so an integrated multi-transit hub was needed. The rebuilding of the Downtown Manhattan transportation network was critical.

Reflections from Carla Bonacci on the Essential Function of the New World Trade Center Transit Hub

The restoration and maintenance of the PATH service between New Jersey and Lower Manhattan was critical, not just for the site but for the recovery of all of Lower Manhattan. Transportation is, in fact, the reason behind everything else.

The PATH Terminal and three subway stations are now directly connected with the WTC Hub, serving hundreds of thousands of daily riders, and setting the stage for all other functions and activities, including the millions of visitors to the WTC site each year.

Yes, there's the commercial part, the component that needed to be met per the legal agreements, and the new retail. The WTC Master Plan allows for flexibility—tenant mixes of large- and smaller-space retailers based on the market. As we continued to develop the Master Plan and guidelines, a very important point was that there could not be any retail facing the Memorial and that retail should be facing Church Street.

The idea of the Master Plan that all of these different uses can live together, that they each can have their own identity, creates a synergy, yet allows the transportation to be a pure function—for commuters and visitors to come and go, to experience the iconic space of the Oculus, and then to partake of the retail.

Isn't this the successful measure of urbanism, the transit-oriented development that we all want? We have it right here.

New Processes and Procedures for Urban Rebuilding

The multitude of stakeholders, each with different objectives, intended to proceed toward satisfaction of those objectives through their usual procedures and formats of decision-making. For example, the Port Authority had established processes by which it handled large-scale urban infrastructure projects, and the communities employed grassroots information programs to gain support and attention to their needs. However, the multi-scaled challenge in this instance made the singular approaches of the various stakeholders problematic and destined for frustration, and this was the most apparent condition of efforts in the early stages of rebuilding.

The challenge was very precisely noted by Alex Garvin[2] as he commenced his role as the Vice President of Design and Planning at the LMDC: "To find a way to implement everyone's agendas—the city government, the state government, the residents, the families."

Without an obvious precedent for such a project and with the heightened social and political sensitivities affecting various dimensions of the project, the activities commenced without an overarching strategic process of combining objectives, coordinating procedures, and cross-referencing progress. This was not an intended approach but rather emerged as many well-intentioned entities rapidly launched their remedial actions. Further, there was an unusual distribution of authority between private ownership and public interests, as well as the interwoven tripartite arrangement of federal, state, and local agencies. The various crises, often vocally discussed by participants and echoed by the media, can be traced predominantly to this strategic shortcoming.

For the rebuilding of Lower Manhattan, the LMDC had been created as a state agency to manage such cooperation of public authorities. However, to gain full functionality, it had to establish a complex governance by combining representatives of the

U.S. government, NYS, New Jersey State, and NYC on its board. Thus, LMDC was able to deal with the following unique challenges:

- The WTC site was owned by a bi-state public authority, the Port Authority, and its planning and land use was therefore not under the jurisdiction of NYC.
- In NYC, some infrastructure systems such as mass transit, electricity distribution, and even the road to the west of the site are under NYS authority. Therefore, NYS, through various agencies, had responsibility for restoring these urban structures and operations.
- The $21 billion of economic aid from the federal government was to be disbursed mostly by federal agencies through suitable state agencies, who then might further fund municipal needs.
- Although a state agency, LMDC was responsible for proposing the allocation of federal funds to specific areas such as the memorialization of lost lives, infrastructure, economic recovery, security, and social rehabilitation.
- And although LMDC was an NYS agency, it shared responsibility with the (bi-state) Port Authority for planning the rebuilding of the latter's WTC site.
- The private sector was to be vitally important, beginning with the critical role of Larry Silverstein, CEO of Silverstein Properties, who, six weeks before 9/11, took a 99-year lease on the entire 16-acre site of the World Trade Center (the Port Authority would retain ownership of the land).

Janno Lieber Reflects

In coming years, we had to deal with two major challenges and two major changes that took place to Silverstein's agreement with the Port Authority, that, in an iterative way, moved us closer to an alignment of interests between the Port Authority and Silverstein and maybe all of the stakeholders as well.

In 2006, we were endeavoring to get some commitments about the schedule of when we would receive the sites for the other three buildings (2 WTC, 3 WTC, and 4 WTC) on the 16-acre site. We were paying hundreds of millions of dollars of rent a year to the Port Authority. And that was diminishing the resources available to rebuild.

There was also the need to clarify the use of the Liberty Bonds, which had come from Washington, DC, and the tax-exempt bonds, which were essential to finance the project. These had become hostage to a recut of the 1 WTC deal, which ultimately led to Silverstein ceding that to the government, although we had started planning for 1 WTC and designed it twice with SOM [Skidmore, Owings & Merrill]. By that time, amazingly, 7 WTC was already completed. It was leasing up. We had confirmed the expectation that Lower Manhattan was going to be a business district again. But there was still uncertainty about when we would get the sites for 2 WTC, 3 WTC, and 4 WTC. So, that 2006 deal created some schedule for the Port Authority and us. This was a good milestone.

There were still some schedule obligations, and ultimately, we worked on a deal with Chris Ward, who was then head of the Port Authority. This was in 2010 when the Port Authority and Silverstein finally teamed up to get 3 WTC and 4 WTC completed.

Silverstein Properties had demonstrated an astounding recovery from the disastrous and tragic scenario of September 11, 2001. Janno Lieber[3] shares some insights on the lessons he learned:

> Here, we did the one thing that I think typifies a successful public/private project, which is that the government, in order to create something of great long-term value, like a building on land owned by the government, provides what it, the government, has to offer to private development, which is to its credit.
>
> Now, the Port Authority had an asset that will pay rent, you know, big rents in the long term. We created alignments of interest. It took much too long, perhaps, but there was another important feature: the commitment and involvement of Community Board 1 and the people who live in the neighborhood became an enormous source of strength for this project, for all of us, and the commitment to seeing the project through. We all remained so aligned in wanting to create a better version of [Downtown] New York than existed even before 9/11.... They [Community Board 1] were great partners after the full airing of differences of opinion and views and micro agendas. Once we accomplished all that and turned it into a plan, it really has been an extraordinary alignment.
>
> And the milestone of the 10th anniversary was huge—for the Port Authority to deliver the Memorial and for people to come walk on the World Trade Center site again. We just can't forget how powerful that was.

Carl Weisbrod Describes the Planning Insights He Gained from Rebuilding Downtown Manhattan

What has guided me in the work I have done since, and certainly the work I'm doing now, is the feeling in Lower Manhattan, pre-9/11, that this was our community. This was a big community (and a sophisticated community), but it was an intimate community in many ways, a community we were all deeply involved in. And all of a sudden, this terrible event happened, and everyone in the world sort of descended on this community and sort of said, "We'll tell you what's best for your community." I've been in government and working with high-level government officials, and I am reasonably sophisticated in dealing with the government. But this spirit of community has affected me.

I've said to myself ever since, and when you're a government official, you've really got to be sensitive to what communities want—it's their community. You can't just go in and say, "We know what's best."

Our efforts became the model for the transformation of older downtowns throughout the country, if not the world, and that had been a joint effort, really, of the Downtown business community and the Downtown residential community.

Passion and Relentless Determination

Even through the technical lessons learned in rebuilding Lower Manhattan, the potential of human interaction and intent to cooperate is notable. There is no managerial formula for repairing urban devastation, but there is a central human capability, as exemplified in so many of those who participated in the rebuilding.

In a review of Lynne Sagalyn's book, *Power at Ground Zero*, Julia Vitullo-Martin concludes her derivation of lessons learned with a focus on the people tending to the rebuilding: "And great leaders make things happen apart from market forces, political dynamics, and institutional context. In the end, with few exceptions, New Yorkers were lucky in the people they had elected to serve them."[4]

In addition, as is so often necessary when the challenge seems too monumental, there is a person or persons who are driven by a determination that is unceasing. For Lower Manhattan, this was Larry Silverstein. He did the first rebuilding at 7 WTC, and he deserves the last word on lessons learned.

Larry Silverstein: Rebuilding the World Trade Center

I was born in Bedford-Stuyvesant, Brooklyn, and lived on the seventh floor of a six-story walk-up! I started with nothing, and with a lot of determination and a little bit of luck, I built a real estate company with some of the smartest, most creative, terrific people from all walks of life.

When I started out, I didn't have a clue as to what to do. Not a clue. And it just came to me that I had to sink or swim. I had to succeed. Failure was not an option.

On the morning of 9/11, I was getting dressed to go Downtown to the World Trade Center. I had just purchased the Twin Towers six weeks before, and I would spend each morning meeting my new tenants. As I was getting ready, my wife, Klara, said, "You can't go! I made an appointment for you with the dermatologist."

I have light hair and fair skin. I have to go to the dermatologist four to five times a year. I said, "But I have so much to do," at which point she got upset and pretty excited. When you're married to someone for 45 years, and they get upset, you have to take it seriously. So I looked at her and I said, "Okay, Okay, don't get

14.3
Larry Silverstein at his desk.
Source: Joe Woolhead, courtesy of Silverstein Properties.

excited. Don't get upset. I'll do whatever you want me to do." She said, "Fine, you're going to the dermatologist." And I said the magic words, "Yes, dear."

And then the call came in, and we saw the tragedy unfold on television. And if it hadn't been for Klara's concern about my well-being, I wouldn't be here today. Simple fact of life.

So, I often tell people that the most important decision you can make in your life is to choose the right spouse or partner. The wrong one is a disaster; the right one enables you to do anything you want to do with your life because you do it together. You do it as partners. You're totally dedicated to each other. You're there for each other without any hesitation or equivocation, without any reservation. It's total, it's a commitment.

I remember sitting with Klara a few days after 9/11 because we had to make a decision. "We can either retire and enjoy our kids and grandkids and travel around the world, or we can spend the next 10 years rebuilding the Trade Center." I said,

"What do you want me to do?" She looked me in the eye and said, "You're not going to be happy doing anything else with your life, so let's get on with rebuilding the Trade Center."

Rebuilding the World Trade Center has clearly been the passion of my life. Of course, it didn't happen overnight, and there were plenty out there who doubted it would ever succeed. Now, you may have heard I'm something of an optimist. It is true I have always been bullish on the future of Lower Manhattan and New York City. But I've been far from alone on that. The progress that we've made is a testament to a great many important voices willing to bet on New York.

After 9/11, people in the architecture, design, and real estate industries poured their hearts and souls into the rebuilding effort. Many of them did inspired work on 7 WTC, the last tower to fall and the first to be rebuilt.

When I started rebuilding 7 WTC, everyone told me I was crazy. All I heard was, "You'll never get it finished. You'll never get it financed. Nobody is going to want to work in a World Trade Center tower building again. Downtown is finished as a business district."

Whether you are in real estate or any other industry, you hear a million different reasons why you can't do what you want to do. Sometimes, the naysayers are going to sound logical and persuasive. So, pretty early on, you have to recognize what's real and what's important and what's white noise or irrelevant. And you cannot allow yourself to lose your focus or be distracted with that stuff. You have to have the courage of your convictions.

We opened 7 WTC five years after 9/11. Designed by my good friend David Childs of Skidmore, Owings & Merrill, it was the first green office tower in New York City history. It is one of the safest buildings ever built and a model not just for all the other World Trade Center buildings but also for the modern American skyscraper.

The building quickly filled up with a diverse group of companies, including branding, technology, and media companies, as well as law and financial services firms. It has clearly been a tremendous architectural, artistic, and commercial success. That success set the stage for the rest of the WTC site and a reborn Downtown.

The story of how New York bounces back, how New York shows resilience, and how New York re-creates itself—also played out next door on the 16-acre WTC site.

After 9/11, there was a great debate over what should be built at the World Trade Center. It was quintessential New York: passionate, loud, and fractious, but one thing was clear—the new World Trade Center needed to be much more than what it had been before.

Of course, we all agreed that our primary responsibility was to commemorate those we lost. At the same time, however, we had to restore the commerce that has defined the lower tip of Manhattan throughout the city's history. Throughout the process, everyone involved was mindful of the delicate balance that had to be achieved.

We began by working with Daniel Libeskind to create a Master Plan that addressed those different goals: It knitted together a fitting and moving tribute to those who lost their lives, a bustling street-level experience with grand open spaces, a train station, as well as new office buildings and stores.

Then, we set about figuring out how to pay for it all. First, I had the obligation and the challenge of collecting on insurance policies with 22 different insurance companies. Now, anyone who has ever suffered a loss knows that filing a claim is only the first step in the process. Well, we litigated for five years to force these insurance companies to pay what they owed so we could rebuild. In the end, it took political pressure from the then-Governor of New York, Eliot Spitzer, who threatened to kick them out of New York if they didn't pay what they owed. So, they finally did.

Sometimes, things take a very long time to do. Rebuilding the World Trade Center has taken 17 years so far, and it will probably be another five years before we are finished. One reason the Trade Center project took so long is because we are a public–private partnership, and the public side kept changing. We were trying to build on a site in New York City that was owned by the Port Authority and overseen by the Governors of New York and New Jersey.

And over the past 17 years, there have been four Governors of New York, six Governors of New Jersey, and three Mayors of New York City! And every time one of them came into office, each with their own agendas, they said, "Stop! Wait! I need to approve this."

There are always going to be people and factors outside your control, but you can't lose focus and you try not to let them slow you down.

After 9/11, I felt a tremendous responsibility, but I wasn't alone. I worked with a team of incredible people: architects, engineers, urban planners, sustainability experts, and even artists. For them—and me—it wasn't just another job. It was a mission to rebuild and create something great out of the devastation of 9/11.

It's important to surround yourself with people who share your values and who are goal-oriented and mission-driven. The people around you need to believe in what you believe in.

At the end of the day, our vision was to create a better version of New York. That vision is now a reality. Today, the new WTC has come alive as a dynamic public space with timeless architecture—and is home to some of the city's most exciting companies.

In 2013, my company opened 4 WTC, the first new building on the original 16-acre site. And with it, we reintroduced a 200-foot stretch of road and sidewalk along Greenwich Street that hadn't existed since the 1960s. Designed by the Pritzker Prize-winning Japanese architect Fumihiko Maki, the building is a model of technology, efficiency, and sustainability, and has also received rave architectural reviews.

We filled up the building with a great group of tenants, including Sports Network, MediaMath, and Morningstar. Interestingly, one of our new tenants is Zurich, an insurance company we fought with after 9/11. They love the building and being Downtown, and our insurance dispute was just part of doing business in the big city.

And last year, we signed a 400,000-square-foot lease with Spotify for the top 14 floors of the tower. It's an amazing tenant for a great building.

We also had a little fun in the building over the few years before we fully leased it up. We reached out to 60 of the city's top street and graffiti artists and invited them to paint on the 69th floor. Well, they took it to heart and painted the walls, floors, ceiling, corridors, elevators, and even the glass. I've always been passionate about public space art, and it was a treat to see what these incredibly talented artists did to this studio in the sky.

The only problem was that it wasn't open to the public—4 WTC is a private office building. So this summer, we took the art outside and commissioned some of those same artists—and some new ones—to paint large murals around the base of 2 WTC. It's a lot of fun watching them paint and seeing how people react to their work.

3 WTC opened in June 2018. The tower was designed by Pritzker Prize-winning architect Richard Rogers. It consists of a reinforced concrete core with a steel structure outside the core and is clad in an external structural steel frame. The building features floor-to-ceiling glass, and all the corners and each floor plate of the tower are column-free to create an open workspace environment—key for cutting-edge companies in brainpower-driven businesses like technology, advertising, and branding, as well as law and financial service firms. Just like 7 WTC and 4 WTC, 3 WTC is a LEED-certified[5] Gold office tower.

The building's anchor tenant is GroupM—the world's largest digital advertising company. They took 700,000 square feet at the base of the tower. McKinsey will move its headquarters into five floors midway up the building, and IEX, the startup stock exchange, will also take a floor. So, the building was nearly 40 percent leased when we opened—not a bad start!

A few years ago, we learned that 21st Century Fox and News Corp would not be moving Downtown into 2 WTC. Of course, that was very disappointing news, but right away, we started a search for a new tenant to anchor 2 WTC—so stay tuned!

This project has always had its twists and turns. It's never been easy, but we never gave up—and the results are visible for the world to see.

Next door, we have 1 WTC, designed by David Childs and recognized by this organization in 2013 as the tallest building in the Western Hemisphere. That building is over 80 percent leased to Condé Nast, High 5 Games, the China Center, some federal agencies, and a number of creative companies. The Observation

Deck at the top of the building is one of the most popular tourist attractions in New York and welcomes millions of visitors a year.

The Santiago Calatrava-designed WTC Transit Hub opened in 2016 and was immediately filled with commuters and visitors. Thanks to the Hub, each of our new buildings will have direct underground access to 11 subway lines and the PATH trains to New Jersey. Westfield's shops are open, including Eataly at the base of 4 WTC. Ultimately, our tenants will have access to 500,000 square feet of destination retail at the WTC and another 350,000 square feet at Brookfield Place, including a brand-new Saks 5th Avenue.

The Memorial opened on the 10th anniversary of 9/11 in September 2011 and to date has welcomed nearly 40 million visitors. The Museum opened in 2014 and has already been visited by over 12 million people.

In addition, the Performing Arts Center unveiled its new design and announced a major gift from Ronald Perelman, which added a vital cultural component to the neighborhood when it opened in 2023. It is home to the Tribeca Film Festival and will host performers and performing arts organizations from all over the world.

Despite all of these great milestones, visible progress all over the site, and tens of thousands of people working there, some people still ask me whether these new buildings will be a success. After all, the American economy still faces uncertainty and challenges.

However, with over 60 years in the business, I've been through a few of these up and down cycles: the 1950s, the oil crisis, the stagflation of the 1970s, and the early 1990s recession. More recently, of course, we've been through the dot-com collapse, the September 11th attacks, the Great Recession following the collapse of Lehman Brothers, and Superstorm Sandy.

From these experiences, I've learned a few things:

First and foremost, I've learned that you should never bet against New York. Never. This town always comes back—stronger than ever.

Second, I learned that if you build it, they will come. You can't always predict how swiftly, but in New York City, first-class office buildings always fill up. And with our city's stock of office buildings aging fast—the average age of a Manhattan office building is 80 years—we need new buildings to protect New York's ability to compete for top-tier jobs.

Third, great companies want to be in high-tech, green buildings. And they want to be right next to the best mass transit and surrounded by dynamic residential neighborhoods filled with brilliant young minds. That's why we've already managed to lease 7.5 million square feet of office space in these new towers alone.

And while we were rebuilding the World Trade Center, Downtown has emerged as a new model of what is best and most dynamic about New York. The population has tripled, and the residential neighborhood around the WTC is now one of the city's most desirable places to live and raise a family.

Downtown isn't your grandfather's Wall Street. While still the international capital of finance, Lower Manhattan has also remade itself into the new media and entertainment capital of America. The area south of Chambers Street now boasts over 800 TAMI (Technology, Advertising, Media, Information) companies, ranging from scores of smaller, ambitious tech startups to media companies like Time Inc., Condé Nast, and HarperCollins to cutting-edge advertising and branding firms like Droga5, GroupM, and MediaMath.

These companies represent the industries that are currently driving the city's new economy—as well as Downtown's future. Their decision to relocate to the World Trade Center proves that New Yorkers were right to bet on Downtown.

It also proves that we made the right policy decision to put jobs and high density next to mass transit: Over the past 10 years, the educated professional workforce has moved to areas in and around Downtown, and we have vastly improved the transportation experience, making it easy for those workers to come to work there. Nearly everyone who works Downtown either walks or takes mass transit to work. That makes Lower Manhattan one of the greenest neighborhoods in the country.

The bottom line is that Downtown New York—a place many had written off—is now the city's hottest neighborhood. The residential population is booming. Incredibly creative people are flocking to our new offices—as well as to our stores and restaurants. New parks and open spaces are coming online seemingly every week. We are adding hotels at breathtaking speed.

Two years ago, we opened our new Four Seasons Hotel, just a block from the WTC, along with Wolfgang Puck's first restaurant in New York. The private residences upstairs at 30 Park Place are doing exceptionally well—the building is now over 85 percent sold. And my wife, Klara, and I moved in earlier this year, so I now have a one-block commute to the office!

I invite you to visit the new World Trade Center and see for yourself what is happening here as we continue to make Downtown Manhattan one of the world's most exciting neighborhoods. Our experience there proves that when passionate New Yorkers work together, we can overcome anything and achieve anything. I think I can speak for all of us when I say how proud I am to have been a part of this. Together, we have reinvented what it means to be a city for the 21st century.

14.4

Statue of Liberty and the annual WTC Light Tribute.

Source: Joe Woolhead, courtesy of Silverstein Properties.

Notes

1 Erika S. Svendsen & Lindsay K. Campbell, "Community-Based Memorials to September 11, 2001: Environmental Stewardship as Memory Work" in K. G. Tidball & M. E. Krasny (eds.), *Greening in the Red Zone: Disaster, Resilience and Community Greening* (Springer Science+ Business Media, 2014), p. 353. DOI 10.1007/978-90-481-9947-1.

2 Alex Garvin, panelist, T+15 Opening Symposium session, December 8, 2016.

3 Janno Lieber, panelist, T+15 Opening Symposium session, December 8, 2016.

4 Julia Vitullo-Martin, "Wrestling Over Hallowed Ground in Downtown Manhattan," *The Wall Street Journal*, September 26, 2016.

5 Certified by the U.S. Green Building Council.

Appendices

Appendix 1
Lower Manhattan T+15 Research Project

LOWER MANHATTAN T+15 is a Columbia University-based, years-long examination of the extraordinary progress made in rebuilding Lower Manhattan in the 15 years since 9/11/2001, with a focus on lessons, insights, and guidelines that can be applied to urban redevelopment projects and city building around the globe. The project was led by Dr. Patrice Derrington, Holliday Associate Professor of Real Estate at Columbia University; Rosemary Scanlon, at the time a Research Scholar at Columbia University's Center for Urban Real Estate; and Robin Panovka, Partner at Wachtell, Lipton, Rosen & Katz and Adjunct Professor, Columbia University.

A series of five seminars were held in 2016 and 2017 to launch research at the Columbia University Center for Urban Real Estate on "T+15," our working title for the project on Rebuilding the World Trade Center, Post 9/11.

Following are the flyers for the sessions.

LOWER MANHATTAN T+15

LESSONS ON URBAN REBUILDING POST 9/11/01

COLUMBIA UNIVERSITY'S CENTER FOR URBAN REAL ESTATE IS PLEASED TO ANNOUNCE:

"LOWER MANHATTAN T+15," an examination of the extraordinary progress made in rebuilding Lower Manhattan in the 15 years since 9/11/2001, with a focus on lessons, insights, and guidelines that can be applied to urban redevelopment projects and city-building around the globe.

A SERIES OF SYMPOSIA ON:

- The Planning Process: The Community and the Public and Private Sectors
- Architectural Insights
- Infrastructure and Transportation
- Building the Residential Community
- Office Space Evolution
- The Public Realm & Memorials in the Cityscape

- Reimagining 24/7 Cities for New Technologies, Pedestrians, and Cyclists
- Urban Economic Revitalization & Other Urban Examples

Recognizing that the world is undergoing massive urbanization and technological change, the project is designed to examine the rebuilding of Lower Manhattan through the lens of proposing progressive, effective guidelines for city-building globally. The project looks forward to the shaping of the future urban environment.

Opening Symposium
December 8, 2016
8:30 to 10:30 a.m.
7 World Trade Center–10th Floor
New York

PROJECT T+15: SCOPE AND GOALS

8:30-8:45 a.m. Patrice A. Derrington, Holliday Associate Professor of Real Estate and Director of the Master of Science in Real Estate Development Program, Columbia University Graduate School of Architecture, Planning and Preservation

Panel 1:

8:45–9:30 a.m. THE PLANNING PROCESS: INVOLVING THE COMMUNITY AND THE PUBLIC AND PRIVATE SECTORS IN FORMULATING THE MASTER PLAN

- **Alex Garvin**, Lower Manhattan Development Corporation
- **Carl Weisbrod**, Chairman, New York City Planning Commission
- **Thomas K. Wright**, President, Regional Plan Association
- **Rosemary Scanlon**, Center for Urban Real Estate Research Scholar, Columbia University **[Moderator]**

Panel 2:

9:30–10:15 a.m. THE PLANNING PROCESS: IMPLEMENTATION AND MODIFICATION OF THE MASTER PLAN

- **Kenneth Lewis**, Partner, Skidmore, Owings & Merrill LLP
- **John N. (Janno) Lieber**, President, World Trade Center Properties, Silverstein Properties, Inc.
- **Christopher O. Ward**, former Executive Director, Port Authority of NY&NJ
- **Robin Panovka**, Partner, Wachtell, Lipton, Rosen & Katz **[Moderator]**

10:15–10:30 a.m. CLOSING REMARKS: LARRY A. SILVERSTEIN, Chairman, Silverstein Properties, Inc.

April 10, 2017
8:30 to 10:00 a.m.
7 World Trade Center–10th Floor
New York

ENVISIONING THE OFFICE MARKET DOWNTOWN, POST 9/11
8:30–10.00 a.m.

- **Janno Lieber**, President, World Trade Center Properties, Silverstein Properties, Inc.
- **Hugh Kelly**, real estate economics consultant and advisor to Fordham University Center for Real Estate
- **Robert von Ancken**, Chairman, Landauer Valuation & Advisory, a division of Newmark Grubb Knight Frank
- **Rosemary Scanlon**, Center for Urban Real Estate Research Scholar, Columbia University **[Moderator]**

May 9, 2017
8:30 to 10:30 a.m.
9/11 Memorial
Museum Entrance at the Museum Pavilion

THE MEMORIALS, THE PUBLIC REALM, AND DOWNTOWN AS A 24-HOUR CITY
Panel 1:
8:30–9:15 a.m.

- **Carla Bonacci**, Assistant Director, WTC Construction, Port Authority of NY&NJ
- **Alex Garvin**, President, AGA Public Realm Strategies, Professor, Yale University,
- **Michael Arad**, Partner, Handel Architects
- **Frank Sciame**, Chairman, F.J. Sciame Construction, Inc.
- **Paul Goldberger [Moderator]**

Panel 2:
9:30–10:15 a.m.

- **Anita Contini**, Arts Program lead, Bloomberg Philanthropies
- **Allison Blais**, Executive Vice President, National September 11 Memorial
- **Hugh Kelly**, author of *24-Hour Cities*, special advisor to Fordham University
- **Charles V. Bagli**, reporter, *Metro NY*, *The New York Times* **[Moderator]**

Appendix 1

May 11, 2017
8:30 to 10:00 a.m.
Offices of Wachtell Lipton
51 West 52nd Street
New York

<u>INFRASTRUCTURE CHALLENGES IN REBUILDING THE WORLD TRADE CENTER AND DOWNTOWN MANHATTAN</u>
8:30–10.00 a.m.

- **Daniel Creighton**, Metropolitan Transportation Authority
- **Tim Gilchrist**, Gilchrist Advisors
- **John Gilbert**, COO of Rudin Management Company
- **Steve Plate**, Chief of Major Capital Projects, Port Authority of NY&NJ
- **Richard T. Anderson,** President Emeritus, New York Building Congress **[Moderator]**

May 18, 2017
8:30 to 10:00 a.m.
The Downtown Alliance
120 Broadway, Suite 3340
New York

<u>DOWNTOWN POST 9/11: REBUILDING THE RESIDENTIAL COMMUNITY</u>
8:30–10.00 a.m.

- **Chris Jones**, Senior Vice President, Chief Planner, Regional Plan Association
- **Carl Weisbrod**, Senior Advisor, HR&A Advisors, Inc.
- **Patrice Derrington,** Director, Center for Urban Real Estate, Columbia University, and The Center for Urban Real Estate **[Moderator]**

Appendix 2
Details of the Unique Public–Private Partnership for the World Trade Center Rebuilding

More in line with other disaster remediation, the rebuilding of Lower Manhattan included public infrastructure and private real estate that had to be interwoven for a seamless urban experience. Typically, public authorities, with their government funding, have attended to restoring the public realm and structures, and private capital, including for the WTC some substantial insurance proceeds, is used to remediate private property. This was the initial approach to reviving Lower Manhattan.

Such an approach to urban recovery has been supported by research that focuses on local issues such as expanded economic resources for victims and increasing resiliency when rebuilding. Other scholars, however, have attended to a larger question: "namely, the broader political context in which strategies of recovery are implemented, and specifically, the debate over the relative merits of government-led vs. 'market-centered' approaches in the recovery effort."[1]

While based on significant political ideologies in the rebuilding of Lower Manhattan, the discussion of who provides the infrastructure and public realm is framed by two unique conditions:

1. Who had the financial resources and rebuilding capability that was so urgently needed?
2. What was the nature of the neighborhood being rebuilt, and did it represent a typical urban community?

In this particular situation, with an estimate of destruction costs of approximately $80 billion in the heart of a leading global city and largely comprising key urban infrastructure and high-density commercial buildings, the challenges of partnerships between government and the private sector were resolved through some surprising modifications.

It is remembered that NYC had a history of "new deal" civic liberalism and large public spending until the recession of the mid-1970s when the budgetary crisis at all levels of government led to pro-business changes to the tax code to attract private funds with Private Activity Bonds (PABs) and Community Development Block Grants (CDBGs). While this approach produced what urban scholars like Harvey have called "urban entrepreneurialism",[2] referred to and further described as a "two-tier econom[y] that exacerbated income inequality and uneven development,"[3] it was firmly in place and utilized in the government's response in the aftermath of September 11, 2001.

Community Development Block Grants and Private Activity Bonds

The legacy of this resourcing structure provided the two main mechanisms for funding the rebuilding of the WTC infrastructure and urban fabric: Community Development Block Grants (CDBGs) and tax-exempt Private Activity Bonds (PABs), which were specifically named "Liberty Bonds." New York State Governor George Pataki and NYC Mayor Michael Bloomberg were given jurisdiction over the $8 billion in bonds, and the LMDC was to oversee $3.7 billion worth of block grants. The Liberty Bond program enabled the NYC Mayor and NYS Governor to allocate up to $8 billion ($4 billion each) in tax-exempt PABs over three and a half years. Ultimately, $6 billion of this was spent.

Historically, tax-exempt PABs had been directed toward privately owned and operated properties that have some public benefit, such as transportation facilities, public works facilities, affordable rental housing, and electric and gas utilities. However, for the WTC rebuilding, the PABs were issued through Congressional legislation in the Victims of Terrorism Tax Relief Act of 2001 and in the Job Creation and Worker Assistance Act of 2002,[4] which removed the requirement for 100 percent public purpose on PABs. Therefore, the tax-exempt Liberty Bonds were made available for *all* rebuilding projects, for public and private benefits, in Lower Manhattan.

The federal government had authorized the CDBG program under Title 1 of the Housing and Community Development Act of 1974. The primary objective of the legislation was the development of viable communities, providing the following, principally for persons of low and moderate income: decent housing, a suitable living environment, and expanded economic opportunities. Funds are distributed and administered by the U.S. Department of Housing and Urban Development (HUD).[5]

As the federal supervising authority, HUD had typically allowed limited waivers to these criteria, often in the name of making it easier and faster for people on low incomes to apply for funds. However, the scale and scope of the waivers granted with respect to the first two criteria for Lower Manhattan were unprecedented. This challenge forced a default on the third criterion applicable to urgent and broader community development needs. Further, in this particular urban area, the conflict had arisen due to the predominant employment needs with respect to the 13 million square feet of office space lost and without any significant low-income residential community.

Additionally, the traditional locational constraints of the PABs were challenged in their implementation in Lower Manhattan. The lifting of caps occurred mainly within the HUD-designated "Liberty Zone." The zone was created using the language of Empowerment Zones—that is, contiguous zones in need of a substantial infusion of investment capital. Yet unlike traditional zones, 25 percent of funds, or $2 billion, could also be spent within NYC at large.

Also unprecedented was the sheer scale of the program. The $8 billion award was almost four times the total amount of all PABs disbursed *nationwide* over a similar period. Most important were the waivers in how PABs could be allocated. First, PABs could be used to spur purely commercial development—in this case, for private real estate and retail property—something never before considered qualified as it did not serve the public interest. Second, traditional "80/20" regulations governing rental housing—with 20 percent of units set aside for "low-income" tenants for 15 years—were now waived. Finally, Liberty Bonds required no official public comment period, giving Pataki and Bloomberg even greater freedom to decide how these massive bonds would be allocated.

In the initial recovery phase, $2.4 billion was allocated, mainly in the form of Business Recovery Grants and Residential Recovery Grants.

For the Business Recovery Grants (BRGs), New York legislators originally sought to target small businesses, by which they envisioned restaurants, retailers, and other small businesses, many of them dependent on the foot traffic that disappeared from Lower Manhattan after the attack. However, as the damage had impacted many larger businesses and commercial entities such as the financial services tenants and commercial landlords in the Lower Manhattan CBD, the NYS agencies—the Empire State Development Corporation (ESDC) and the Lower Manhattan Development Corporation (LMDC)—defined such businesses broadly as any company with fewer than 500 employees and with no restrictions on annual revenues. As a result, close to 40 percent of the BRGs supported major corporations with hundreds of employees, as well as financial trading and legal partnerships.

Private-Sector Funding

Although the government is traditionally responsible for the nation's defense, technology has increased the reliance on independent telecommunications and connection for military management and control, and also for critical infrastructure such as power plants and electrical grids, gas and oil distribution pipelines, water treatment systems, chemical refineries, and transportation systems. The Clinton administration created the Critical Infrastructure Assurance Office (CIAO) within the Department of Commerce to coordinate the federal government's initiatives on critical infrastructure assurance efforts with the private sector and to support the Information Sharing and Analytics Centers (ISACs), which were established for various sectors to promote collaboration between all levels of government and industry.

As is usually important in public–private arrangements, the sharing of information requires careful handling, and both the Clinton and G. W. Bush administrations emphasized that the ISACs were not to interfere with routine information exchanges between the government and industry, but rather they were to have a technical focus with respect to non-regulatory and non-law enforcement missions. As such, they became a clearinghouse for information within and among members and across sectors.

Other examinations of private funding for infrastructure have noted the complexity of responsibility when those structures and operations are regarded as critical to national security. When serving as Assistant Secretary of Commerce (1993-1997), Sue Eckert[6] pressed for engagement of the private sector in creating and maintaining critical infrastructure. In her white paper (16 July, 2002), she pointed out that the destruction and rebuilding of the WTC prompted motivation for both government and industry to pursue cooperative infrastructure mechanisms that had previously languished. She further recognized the significance of the recent initiatives of the Information Sharing and Analytics Centers (ISACs),[7] which were established for various sectors to promote collaboration between all levels of government and industry.

Remaining problematic for the rebuilding in Lower Manhattan, however, was the high cost of implementing increased security measures directed by the information and technical standards provided by the ISACs. The Congressional 9/11 Commission's Report[8] had warned that the lack of a widely embraced private-sector preparedness standard was a principal contributing factor to the lack of preparedness for defending America from attack.

Bringing together the public and private resources for the rebuilding of Lower Manhattan provides an exceptional and instructive example of how such an important collaboration is possible. The achievements over these past 20+ years, though sometimes struggling with the forced intersection of interests, could not have been realized without such a dynamic.

Notes

1 K. Fox Gotham & M. Greenberg, "From 9/11 to 8/29: Post-Disaster Recovery and Rebuilding in New York and New Orleans," *Social Forces* 87(2) (December 2008), pp. 1039–1062; quotation, p. 1040.

2 David Harvey, "From Managerialism to Entrepreneurialism: The Transformation in Urban Governance in Late Capitalism," *Geografiska Annaler. Series B, Human Geography* 71(1) (1989), pp. 3–17; quotation, p. 4.

3 Fox Gotham & Greenberg, "From 9/11 to 8/29" (2008), p. 1044.

4 H.R.2884—107th Congress (2001-2002) - Victims of Terrorism Tax Relief Act of 2001; H.R.3090—107th Congress (2001-2002) - Job Creation and Worker Assistance Act of 2002.

5 Further information can be found at www.hud.gov/sites/documents/DOC_16470.PDF.

6 Sue Eckert, *The National Strategy for Homeland Security* (Office of Homeland Security, 16 July 2002); www.whitehouse.gov/homeland/book/nat_strat_hls.pdf. The author's views on this issue were informed by her role as Assistant Secretary of Commerce from 1993 to 1997, when she was involved with early efforts by the Clinton administration to engage the private sector on critical infrastructure issues.

7 For further information, see www.nationalisacs.org/.

8 National Commission on Terrorist Attacks upon the United States, public report (2004); see www.9-11commission.gov.

Appendix 3
Timetable of World Trade Center Projects

Project	*Completion Date*
7 WTC	May 2006
September 11 Memorial	September 2011
4 WTC	November 2013
September 11 Museum	May 2014
1 WTC	November 2014
WTC Oculus Transit Hub	March 2016
Liberty Park	June 2016
Westfield WTC Retail Complex	August 2016
3 WTC	June 2018
WTC Cortlandt Street Subway Station	September 2018
St. Nicholas Greek Orthodox Church	December 2022
Perelman Performing Arts Center (PAC)	September 2023
2 WTC	On hold
5 WTC	On hold

Appendix 4
Biographies of Contributors and Writers of Reflections (in alphabetical order)

Robert von Ancken

Robert ("Bob") von Ancken is a highly respected real estate appraiser, beginning his career with Grubb and Ellis in 1985 and later as President of Landauer Valuation Company. In 1998, Landauer was purchased by Newmark Knight Frank; Bob served as head of their new appraisal operations until 2018, when he "semi-retired." Over his career, he has appraised the value of more than 6,000 commercial properties, primarily office buildings.

Mr. von Ancken was President of the New York Metro Chapter of the Appraisal Institute, he is a designated member of the Appraisal Institute (MAI), and has been a member of the Counselors of Real Estate (CRE) since 1981. In his spare time, he has renovated a Manhattan landmark townhouse, homes in East Hampton, LI, and a hotel in Southern Vermont, among others. In semi-retirement, he manages 10 commercial properties.

Figure A4.1
Robert von Ancken.

Richard T. Anderson

Richard T. Anderson ("Dick") is President of RTA Advisory Services LLC, a private firm specializing in government relations and strategic industry planning. He is President Emeritus of the New York Building Congress (NYBC), which he led for more than 20 years, ushering in a period of influence and growth for this public policy organization for the construction and real estate communities.

Prior to his years at the NYBC, Mr. Anderson served as Executive Director of the Dallas Plan, preparing a long-range capital improvements strategy for that city, and was President of the Regional Plan Association, the nation's oldest metropolitan planning organization serving the tri-state New York metropolitan area. Mr. Anderson has served as President of the American Planning Association and is a founding member of the American Institute of Certified Planners, among other leadership positions. He was awarded the 2001 George S. Lewis Award from the New York Chapter of the American Institute of Architects (AIA), and the Building Congress Board of Directors presented him with their Distinguished Service Award in 2004.

Figure A4.2
Richard T. Anderson.

Richard Anderson is a graduate of Rutgers University and earned a Master of Regional Planning degree from Cornell University. He has completed postgraduate doctoral studies in public administration at New York University.

Michael Arad

Figure A4.3
Michael Arad.

Michael Arad's design for the National September 11 Memorial at the World Trade Center site, titled "Reflecting Absence," was selected by the Lower Manhattan Development Corporation from among more than 5,000 entries submitted in an international competition held in 2003. Mr. Arad joined Handel Architects as a Partner in April 2004, where he worked on realizing the Memorial design as a member of the firm.

A native of Israel, Mr. Arad was raised there, in the UK, the United States, and Mexico. He came to the United States and earned a BA from Dartmouth College in 1994 and a Master of Architecture degree from the Georgia Institute of Technology in 1999. Mr. Arad became a resident of NYC following his studies. He worked for Kohn Pedersen Fox in the city before joining the Design Department of the New York City Housing Authority, where he was working during the Memorial competition.

In 2006, Mr. Arad was one of six recipients of the Young Architects Award of the American Institute of Architects (AIA). In 2012, he was awarded the AIA Presidential Citation for his work on the National September 11 Memorial. In addition, he was also honored in 2012 by the Lower Manhattan Cultural Council with the Liberty Award for Artistic Leadership. His work on the Memorial was recognized with Honor Awards from the AIA and the American Society of Landscape Architects (ASLA).

In 2017, Mr. Arad was selected to design a memorial to the victims of the 2015 Charleston church massacre at the Mother Emanuel AME Church in Charleston, South Carolina.

Charles Bagli

Figure A4.4
Charles Bagli.

Journalist Charles Bagli is the author of *Other People's Money: Inside the Housing Crisis and the Demise of the Greatest Real Estate Deal Ever Made* (Dutton/Penguin, 2013). For 21 years, he wrote for *The New York Times*, covering the intersection of politics and real estate. He reported stories about the rebuilding of Lower Manhattan, the rise of Billionaires' Row, public investment in casinos and stadiums, and the rise and fall of the $6 billion Stuyvesant Town sale.

In his earlier career, he worked for the *New York Observer* and the *Tampa Tribune*.

Allison Blais

Figure A4.5
Allison Blais.

Allison Blais is the Executive Vice President and Chief Strategy and Operations Officer for the National September 11 Memorial & Museum (also known as the 9/11 Memorial Museum). She is also co-author of *A Place of Remembrance* (2011), the official book of the organization. Ms. Blais is responsible for managing priority institution-wide initiatives and oversees a team focused on the 9/11 Memorial Museum's ongoing sustainability and relevance, as well as its daily operations.

Early in the replanning effort, Ms. Blais worked in the Chairman's and President's offices of the LMDC. She then continued at the National September 11 Memorial & Museum, where she served as the organization's Chief of Staff for eight years through the design, content development, building, opening, and operational ramp-up of the Memorial in 2011 and the Museum in 2014. Since the Museum's opening, she has served in many roles, including Chief Operating Officer and, currently, Chief Strategy and Operations Officer.

Before her work Downtown, Ms. Blais managed capital projects for major cultural institutions and construction companies throughout NYC and worked at the Public Theater/New York Shakespeare Festival. She graduated *summa cum laude* from Cornell University and received a Master's degree from Columbia University. She lives in Lower Manhattan with her husband and daughter.

Carla Bonacci

Figure A4.6
Carla Bonacci.

For the past 33 years, Carla Bonacci has led architecture/engineering/construction (AEC) teams to plan, design, construct, and regulate public and private projects at the World Trade Center. Her tenure and work have spanned the modernization of the WTC complex after the 1993 terrorist bombings, the restoration of the WTC site after the 9/11 terrorist attacks, and the redevelopment and ongoing rebuilding of the WTC Campus. In the aftermath of 9/11, Ms. Bonacci provided critical leadership to assert public objectives and institute the planning principles that are the basis of the WTC Campus Master Plan.

Ms. Bonacci has applied her expertise across a range of capital projects (both pre- and post-9/11) to redevelop public spaces, construct central plant facilities, and install site-wide life safety and security systems, as well as providing project oversight of third-party base building and interior fit-out projects through occupancy. She organized and led teams on the WTC Campus Security Program and the WTC technology systems and, following the damage caused by Superstorm Sandy in 2012, led a team of over 100 individuals to prepare the adjacent World Financial Center (WFC) Flood Mitigation Program.

Her recent work includes integrated projects and oversight supporting the completion of the Perelman Performing Arts Center and the St. Nicholas Greek Orthodox Church, involving the completion of the streetscape that defines the WTC Campus, installations of deployable flood mitigation barriers that preserve the WTC's aesthetic quality, and the final phase of WTC's Liberty Park with a Memorial Garden commemorating Port Authority first responders—each of these receiving industry recognition and awards.

Ms. Bonacci dedicates her professional career as well as her prestigious Fellow of the American Institute of Architects (FAIA) designation in remembrance of the World Trade Center co-workers and Port Authority colleagues lost in February 1993 and September 2001, and she knows that they would be proud of what has been accomplished.

Figure A4.7
Ric Clark.

Ric Clark

Ric Clark is Co-Founder and Managing Partner of WatermanCLARK, a vertically integrated real estate investment and operating company that was formed in 2020.

Prior to WatermanCLARK, Mr. Clark had been with Brookfield Asset Management and its affiliates since 1984 in various senior roles, including Chairman and Chief Executive Officer of Brookfield Property Group, Brookfield Property Partners, and Brookfield Office Properties. Under his leadership, the property group grew its assets under management from less than $5 billion to over $200 billion. It expanded its investment, operational, and asset management capabilities beyond the office sector into the multifamily, industrial, hotel, and retail sectors; and extended its geographic footprint beyond North America and Brazil to Europe, Australia, India, and China. Today, Brookfield is one of the largest owners and operators of real estate, with more than 600 million square feet of properties in key markets worldwide.

Mr. Clark is Chairman of the Alliance for Downtown New York and the Downtown-Lower Manhattan Association. He serves on the executive committees of the Real Estate Board of New York, the Real Estate Roundtable, the board of trustees of the National September 11 Memorial & Museum and the Ronald O. Perelman Performing Arts Center at the World Trade Center. Mr. Clark holds a Bachelor of Science in business from the Indiana University of Pennsylvania.

Figure A4.8
Natalie R. Cohen.

Natalie R. Cohen

Natalie R. Cohen is the Founder of National Municipal Research, an independent consulting and research company focused on U.S. state and local government finance. She publishes *The Public Purse*. A seasoned analyst, writer, and speaker, she started her career at New York City's Office of Management and Budget (OMB). She has worked at a rating agency in bond insurance, and is known for her published work on municipal bond defaults.

Ms. Cohen was recognized by the *Bond Buyer* as a "Trailblazer" and recently received the Career Achievement Award from the National Federation of Municipal Analysts (NFMA). Ms. Cohen sits on the board of Build America Mutual, a bond insurer, and chairs the Audit Committee. Ms. Cohen also advises Riskbridge, an "outsourced chief investment officer (OCIO)," that advises not-for-profit entities, pension plans, insurance companies, and family offices. In addition, Ms. Cohen is a Senior Fellow with the Milken Institute Center for Financial Markets. She holds an undergraduate degree from Hampshire College and a Master's degree from NYU Graduate School of Public Administration (now called the Wagner Graduate School of Public Service).

Anita Contini

Figure A4.9
Anita Contini.

Anita Contini has been the program lead of arts and culture for Bloomberg Philanthropies since October 2010. From 2005 to 2010, she was Senior Vice President and Director of Corporate and Public Affairs for the CIT Group.

Between August 2002 and July 2005, Ms. Contini was Vice President and Director of Memorial and Cultural Programs at the Lower Manhattan Development Corporation, where she led the important worldwide competition for the design of the WTC Memorial. In earlier years, she initiated a series of cultural events for Downtown Manhattan.

Paul Goldberger, in his 2004 book *Up from Zero* (Random House), described her as "widely respected by both established, uptown arts institutions and by younger, more cutting-edge arts groups and artists." And "she was comfortable working within a large organization" (p. 214) based on her background as Vice President and Artistic Director of Brookfield Properties/Olympia & York and later as an executive at Merrill Lynch, where her portfolio included global sponsorships and events marketing.

Ms. Contini is a graduate of Hofstra University and lives in NYC.

Daniel Creighton

Figure A4.10
Daniel Creighton.

Daniel Creighton is the Senior Vice President and Business Unit Leader for Signals and Train Control at MTA Capital Construction. He has been with the MTA since 2016, and among other notable achievements, he was responsible for the design, delivery, and completion of the $181.8 million Cortlandt Street #1 Subway-line station in the middle of the World Trade Center site. This was completed with the coordination of Port Authority, NYC Transit Authority, MTA personnel, private developers, contractors, and design consultants. He opened the WTC Cortlandt Station below budget and almost one year ahead of the project schedule in 2018.

Prior to his career at the MTA, Mr. Creighton served with the U.S. Department of Transportation, Federal Transit Administration in New York from 2005 to 2016; he was Director Megaprojects for the Lower Manhattan Recovery Office (2013–2016); team leader/lead engineer, Lower Manhattan Recovery Office (2009–2013); and initially senior engineer, Lower Manhattan Recovery Office (2005–2009). During these years, Daniel was responsible for administering nearly $5 billion in emergency federal aid to the New York region following the terrorist attacks of September 11, 2001, as well as a portfolio of $15 billion for megaprojects, including East Side Access and the Second Avenue Subway.

Mr. Creighton holds a Bachelor of Science in Civil Engineering (B.S.CivE) and a Bachelor of Science in engineering and public policy from Carnegie Mellon University (1992). He earned a Master of Business Administration (MBA) degree from Baruch College, City University of New York (CUNY) (2005), with a major in finance and investments.

Figure A4.11
Alex Garvin.

Alex Garvin

Alexander ("Alex") Garvin passed away in December of 2021 at the age of 80 in Manhattan, where he was born and lived all his life. He combined a career in urban planning and real estate with teaching, architecture, and public service. He was President and CEO of AGA Public Realm Strategists, Inc., a planning and design firm in NYC that was responsible for initial Master Plans for the Atlanta BeltLine; Tessera (a 700-acre new community outside Austin, Texas); and Hinton Park in Collierville, Tennessee. Between 1996 and 2005, he was Managing Director for Planning at NYC2012, the committee established to bring the Summer Olympics to New York in 2012.

During 2002–2003, Alex Garvin was Vice President for Planning, Design, and Development, responsible for planning the rebuilding of the World Trade Center for the Lower Manhattan Development Corporation (LMDC). He was largely responsible for bringing on Daniel Libeskind as the Master Planner of the WTC site.

Over 46 years, he held prominent positions in five NYC administrations, including Deputy Commissioner of Housing and City Planning Commissioner.

In addition to his professional work, for the past 50 years, Mr. Garvin taught at Yale University, his *alma mater*, where, as an Adjunct Professor of Urban Planning and Management, he taught a wide range of courses in architecture, city planning, and real estate development. He also taught workshops on basic real estate development for the Urban Land Institute.

In his obituary for *The New York Times* on December 21, 2021, Paul Goldberger wrote:

> He wrote several books on cities, including *The American City: What Works, What Doesn't*, first published in 1996. That book's blunt title was typical of Mr. Garvin, who had no patience with urban planning dogmas that reduced cities to all-encompassing formulas. To him, cities were more than physical plans, and their forms inevitably resulted from myriad political, social, economic, and aesthetic forces. It was the planner's job, he believed, to manage and guide these forces and to target crucial public investments that would spur private development. "The public realm," Garvin said, "is the framework around which everything else grows."

Alex Garvin authored *The American City: What Works and What Doesn't* (McGraw-Hill, 1996) now in its third edition; *The Planning Game, Public Parks: The Key to Livable Communities* (Norton Professional Books, 2010), and his newest book, *What Makes a Great City* (Island Press, 2021).

Mr. Garvin won numerous awards, including the Municipal Art Society's New York City Masterwork Award for Best Planning and Urban Design, the AIA (New York Chapter) Merit Award, and the American Planning Association (NYC Chapter) Distinguished Service Award.

John Gilbert

John Gilbert is Co-Founder and Executive Chairman at Prescriptive Data, an award-winning smart building tech firm whose flagship building operating system and energy management solution, Nantum OS, connects built environments while optimizing energy efficiency and tenant comfort and wellness. Nationally acknowledged as an industry thought leader in the integration of technology into the built environment and a frequent public speaker, John's vision is to "future-proof real estate by giving all buildings a brain."

Mr. Gilbert was previously COO/Executive Vice President/Chief Technical Officer of Rudin Management Company (RMC), the largest privately owned real estate company in NYC, which manages over 15 million square feet of commercial and residential space. He has been involved in every major project the company has developed over the last 25 years, including the redevelopment and creation of the world's first smart building at 55 Broad Street (1995–1996), the development of the Reuters Building at 3 Times Square (1998–2001), the redevelopment of the former ATT Long Lines Building at 32 Avenue of the Americas (1999–2002), the redevelopment of 130 West 12th Street (2010–2011), the development of The Greenwich Lane Condominiums (2006–2016), and Dock 72, a ground-up 675,000 square-foot office building located in the Brooklyn Navy Yard, a joint venture with WeWork and Boston Properties.

Figure A4.12
John Gilbert.

Tim Gilchrist

Tim Gilchrist has over 32 years of experience in government. He is currently retired and serves as a Lecturer in public policy at Saint Peter's University, NJ, and as President of Gilchrist Advisors. Prior to retirement, he served as President of the Moynihan Station Development Corporation, responsible for developing NYC's historic Farley Post Office into Amtrak's new intercity station. During Mr. Gilchrist's tenure, the long-planned and debated project moved into the construction phase, with the first projects starting construction in the fall of 2012.

Previously, Mr. Gilchrist served as the Senior Advisor for Infrastructure and Transportation and Deputy Secretary for Economic Development and Infrastructure for NYS Governors Spitzer and Paterson, responsible for oversight of the activities of NYS's transportation, economic development, housing, gaming, and arts agencies. He was the lead for Governor Paterson on the World Trade Center site, including the negotiations that resulted in the 2010 financing agreement. Tim also served as the Chairman of the New York State Economic Recovery and Reinvestment Cabinet, managing and overseeing over $28 billion in New York State's American Recovery and Reinvestment Act (ARRA) funds.

Prior to serving in the Governor's Office, Mr. Gilchrist held multiple high-level management positions in the New York State Department of Transportation, including overseeing the rebuilding of West Street from the Battery to Chambers Street as part of the redevelopment of Lower Manhattan. He holds a Master of Public Administration degree from the Harvard Kennedy School of Government and a BA from the State University of New York (SUNY) at Oswego.

Figure A4.13
Tim Gilchrist.

Figure A4.14
Paul Goldberger.

Paul Goldberger

Paul Goldberger, whom the *Huffington Post* has called "the leading figure in architecture criticism," is a Contributing Editor at *Vanity Fair*, and from 1997 through 2011, served as Architecture Critic for *The New Yorker*, where he wrote the magazine's celebrated "Sky Line" column. He also holds the Joseph Urban Chair in Design and Architecture at The New School in NYC. He was formerly Dean of the Parsons School of Design, a division of The New School. He began his career at *The New York Times*, where, in 1984, his architecture criticism was awarded the Pulitzer Prize for Distinguished Criticism, the highest award in journalism.

Mr. Goldberger is the author of several books, including *Ballpark: Baseball in the American City* (Alfred A Knopf, 2019). His other works include *DUMBO: The Making of a Neighborhood and the Rebirth of Brooklyn* (Rizzoli, 2021); *Blue Dream and the Legacy of Modernism in the Hamptons* (Delmonico Books, 2023); *Building Art: The Life and Work of Frank* Gehry (Alfred A. Knopf, 2015); *Why Architecture Matters* (Yale University Press, 2009; new edition, 2023); and *Building Up and Tearing Down: Reflections on the Age of Architecture*, a collection of his architecture essays (Monacelli Press, 2009).

Paul Goldberger's chronicle of the process of rebuilding Ground Zero, entitled *Up from Zero: Politics, Architecture, and the Rebuilding of New York*, published by Random House in the fall of 2004 and re-released in an updated paperback edition in 2005, was named one of The New York Times Notable Books for 2004. Mr. Goldberger has also written *The City Observed: New York, A Guide to the Architecture of Manhattan*; *The Skyscraper; On the Rise: Architecture and Design in a Post-Modern Age; Above New York;* and *The World Trade Center Remembered*.

He lectures widely around the country on the subjects of architecture, design, historic preservation, and cities, and he has taught at both the Yale School of Architecture and the Graduate School of Journalism at the University of California, Berkeley, in addition to The New School. His writing has received numerous awards in addition to the Pulitzer, including the President's Medal of the Municipal Art Society of New York, the Medal of the American Institute of Architects, and the Medal of Honor of the New York Landmarks Preservation Foundation, awarded in recognition of "the nation's most balanced, penetrating and poetic analyses of architecture and design." He is a graduate of Yale University.

Figure A4.15
T. J. Gottesdiener.

T. J. Gottesdiener

T. J. Gottesdiener is an architect and was the Managing Partner of the New York office of Skidmore, Owings & Merrill (SOM). A graduate of Cooper Union's Irwin S. Chanin School of Architecture, Mr. Gottesdiener joined SOM in 1980 and was made Partner in 1994. In 2023, he joined The Georgetown Company to work as an architect-developer.

Mr. Gottesdiener is committed to enhancing the built environment of NYC and was responsible for some of SOM's most complex and challenging projects in Manhattan. Closely involved in the revitalization of Lower Manhattan and the redevelopment of the WTC site, he played a major role in the planning and phasing for the design and construction of 13 million square feet of commercial construction in cooperation with agencies responsible for the memorial, cultural, and transportation functions. He was Managing Partner for 7 WTC, which was completed in 2006, and 1 WTC, which was completed in July 2013.

Mr. Gottesdenier is a Fellow of the American Institute of Architects. He lives with his wife in NYC; they have one son.

Alice M. Greenwald

Alice M. Greenwald is the Founder and Principal of Memory Matters, LLC, providing strategic advising services for museums, memorial projects, senior executives, and boards, with a focus on helping communities build resilience through the commemoration and documentation of traumatic history.

From January 2017 to September 2022, Ms. Greenwald served as President and CEO of the National September 11 Memorial & Museum, where she was responsible for the overall vision, financial well-being, management, and long-term sustainability and relevance of the 9/11 Memorial & Museum. From 2006–2016, Ms. Greenwald was Executive Vice President for Exhibitions, Collections, and Education and Director of the 9/11 Memorial & Museum. In this role, she oversaw the articulation and implementation of a founding vision for the 9/11 Memorial & Museum, managing its programming, collecting, exhibition, and educational initiatives.

Figure A4.16
Alice M. Greenwald.

Ms. Greenwald previously served as Associate Museum Director, Museum Programs at the United States Holocaust Memorial Museum. Her 19-year affiliation with USHMM began in 1986 when she served as a member of the "design team" for the permanent exhibition.

Her book, *The Stories They Tell: Artifacts from the National September 11 Memorial Museum*, co-edited with Clifford Chanin and published by SkiraRizzoli, was cited in *The New York Times* as one of the best books about NYC published in 2013. She is also the Executive Editor and primary contributing author of *No Day Shall Erase You: The Story of 9/11 as Told at the National September 11 Memorial Museum* (SkiraRizzoli, 2016).

Ms. Greenwald serves on the boards of the United Kingdom Holocaust Memorial Foundation, the International Council of Museums (U.S.), and the Central Synagogue in NYC. She is the first Vice President of The Lotos Club. She holds an MA in the history of religions from the University of Chicago Divinity School and a BA with concentrations in English literature and anthropology from Sarah Lawrence College, where she delivered the commencement address to the class of 2007.

Christopher Jones

As a Senior Research Fellow at the Regional Plan Association (RPA), Chris Jones's work focuses on how urban policy shapes metropolitan economies and their social, natural, and built environments. In his previous role as RPA's Senior Vice President and Chief Planner, he directed economic, housing, transportation, and environmental research and planning. He led the research and policy development for RPA's Fourth Regional Plan, a long-range blueprint released in 2017 for tackling systemic inequality and climate change and modernizing the New York-New Jersey-Connecticut region's infrastructure and institutions to promote broadly shared health and prosperity. Most recently, his work has focused on how the tri-state region's economy and future development will be shaped by changes unleashed by the Covid-19 pandemic.

Figure A4.17
Christopher Jones.

Since joining RPA in 1994, he has led multidisciplinary initiatives and analyses to improve economic, social, and environmental conditions in the tri-state region, including a federal Sustainable Communities initiative that included 17 cities, counties, and planning organizations; a series of reports on the future of Long Island; alternatives for Manhattan's Far West Side; plans for transit-oriented and community development throughout the region; and the economic and workforce development chapters of *A Region At Risk*, RPA's Third Regional Plan.

Prior to joining RPA, Mr. Jones was the Special Assistant to the Deputy Mayor for Planning and Development in New York City and Supervisor of Forecasting for the Port Authority of New York and New Jersey. He has a Master of Urban Planning degree from Hunter College of the City University of New York and a bachelor's degree from the State University of New York at Albany.

Figure A4.18
Hugh F. Kelly.

Hugh F. Kelly

Hugh Kelly heads his own consulting practice, Hugh F. Kelly Real Estate Economics, which serves the private sector, governmental organizations, and not-for-profit agencies.

Mr. Kelly served as the Board Chairman of the Counselors of Real Estate in 2014 and, in November 2019, received the CRE's highest honor, the Landauer/White Award, for his professional achievements and his civic and nonprofit community work. He served six years as Board President of the affordable housing agency of Brooklyn/Queens Catholic Charities and is presently a Vice Chair of the diocese's real assets management board.

Hugh Kelly's book, *24-Hour Cities: Real Investment Performance, Not Just Promises* (Routledge, 2016), won the Gold Award for books from the National Association of Real Estate Editors in 2017. He is co-author of a forthcoming textbook on real estate capital markets.

Mr. Kelly was the lead author of *Emerging Trends in Real Estate* from 2015 to 2020 and of Integra Realty Resources' *Viewpoint* from 2017 to the present. His columns on economics and real estate appear regularly in *Commercial Property Executive* and *The Appraisal Journal*. He earned his BA in philosophy (*magna cum laude*) from Cathedral College (Douglaston, New York) and his PhD in real estate and the built environment at the University of Ulster (Belfast, Northern Ireland).

Hugh Kelly taught from 1984 to 2016 at NYU's Schack Institute of Real Estate, where he earned the rank of full Clinical Professor. From 2017 to 2021, he helped Fordham University establish its Master's degree program in real estate.

Figure A4.19
Kenneth Lewis.

Kenneth Lewis

Kenneth Lewis joined Skidmore, Owings & Merrill (SOM) in 1986, and is now a Partner at the firm. His more than 35-year career includes 1 WTC and its neighbor, 7 WTC, the first to rise at the site after 9/11 and the first LEED Gold skyscraper. His recent projects include 35 Hudson Yards, an undulating residential and office tower, and Manhattan West, a 7 million square foot, mixed-use, Master-Planned development built above active railroad tracks.

Mr. Lewis is also at the forefront of another transformation, the sustained focus on climate change and the need for equity in the community and the profession. At SOM, he directed resources to work for affordable housing at New Jersey Performing Arts Center (NJPAC) and the Lower Eastside Girls Club of New York annex. He currently serves on the boards of directors for AIA New York, AIA Architecture Foundation, and the Urban Green Council. As the 2021 President of AIA New York, Mr. Lewis was charged with leading the organization through one of the AEC (architecture, engineering, and construction) industry's most challenging periods in recent history. During his leadership, the chapter, in collaboration with NYCOBA (NY Chapter of the National Organization of Minority Architects), launched the AIANY 2030 Fund, which provides both debt relief and mentorship support during the licensing process for young BIPOC (Black, Indigenous, People of Color) architects. Mr. Lewis holds a BA in art & architecture from the Rhode Island School of Design, and an MA in architecture from the Harvard University Graduate School of Design.

Daniel Libeskind

Figure A4.20
Daniel Libeskind.

An international figure in architecture and urban design, Daniel Libeskind is renowned for his ability to evoke cultural memory in buildings. Informed by a deep commitment to music, philosophy, literature, and poetry, Mr. Libeskind aims to create architecture that is resonant, unique, and sustainable.

Born in Łód, Poland, in 1946, Mr. Libeskind immigrated to the United States as a teenager and, with his family, settled in the Bronx. He received the America-Israel Cultural Foundation Scholarship and performed as a musical virtuoso before eventually leaving music to study architecture. He received his professional degree in architecture from the Cooper Union for the Advancement of Science and Art in 1970 and a postgraduate degree in the history and theory of architecture from the School of Comparative Studies at the University of Essex in England in 1972.

In 1989, Mr. Libeskind won the international competition to build the Jewish Museum in Berlin. He moved his young family to Berlin and devoted more than a decade to the completion of this seminal design. A series of influential museum commissions followed, including the Felix Nussbaum Haus, Osnabrück; Imperial War Museum North, Manchester; Denver Art Museum; Contemporary Jewish Museum, San Francisco; Danish Jewish Museum, Copenhagen; Royal Ontario Museum, Toronto; and the Military History Museum, Dresden.

In 2003, Studio Libeskind won another historic competition—to create a Master Plan for the rebuilding of the World Trade Center in Lower Manhattan. In addition to a towering spire of 1,776 feet, the Libeskind design study proposed a complex program encompassing a memorial, an underground museum, the integration of the slurry wall, a special transit hub, and four office towers. This plan is being realized today.

Upon moving to New York, Studio Libeskind quickly became involved with designing and realizing a large number of commercial centers, such as Westside in Bern, the Crystals at City Center in Las Vegas, and Kö-Bogen in Düsseldorf, as well as residential towers in Busan, Singapore, Warsaw, Toronto, Manila, and São Paulo.

As Principal Design Architect for Studio Libeskind, Mr. Libeskind speaks widely on the art of architecture at universities. His architecture and ideas have been the subject of many articles and exhibitions, influencing the field of architecture and the development of cities and culture.

Mr. Libeskind lives in New York with his wife and business partner, Nina Libeskind. He is a licensed architect in the State of New York.

Figure A4.21
Janno Lieber.

Janno Lieber

John ("Janno") Lieber was confirmed as Chairman and CEO of the Metropolitan Transportation Authority (MTA) in January 2022, following his nomination to that position by Governor Kathy Hochul. Mr. Lieber has been serving as MTA Acting Chairman and CEO since July 2021, and is the Authority's 15th Chairman since it was founded in 1968.

In this position, Mr. Lieber provides strategic, financial, and operational leadership for the MTA, the largest public transportation agency in North America. He also heads the MTA Board of Directors, senior leadership team, and a total workforce of more than 60,000 employees.

From 2017 to 2021, Mr. Lieber served as President of MTA Construction & Development, the organization responsible for planning, developing, and delivering capital projects across the MTA. In that position, he led the creation and implementation of the MTA's 2020–2024 $51.5 billion capital program. He also managed the integration and transformation of construction and development management practices at the MTA.

Prior to the MTA, Janno Lieber served as President of World Trade Center Properties LLC, where he was responsible for managing all aspects of the Silverstein organization's efforts to rebuild the World Trade Center site, including planning, design, and construction issues; business, financing, and legal matters; and public affairs, government, and community relations.

Earlier in his career, Mr. Lieber held positions in the administrations of President Bill Clinton and NYC Mayor Ed Koch and worked as an attorney in private practice. He holds a BA from Harvard University and a JD in law from the New York University School of Law.

Figure A4.22
Eugene McGrath.

Eugene ("Gene") R. McGrath

Gene McGrath was the Chairman of the Board and Chief Executive Officer of Consolidated Edison Company of New York, Inc. (ConEd), a regulated utility providing electric, gas, and steam service in NYC and Westchester County, New York. A native New Yorker, born in 1942 in Yonkers, Mr. McGrath spent his entire professional career with ConEd.

Mr. McGrath graduated from Manhattan College in 1963 with a Bachelor's degree in mechanical engineering. He accepted a position with ConEd directly upon graduation. He started his career as an engineer and management trainee in the operating and customer-service divisions and continued to move progressively into more responsible positions, eventually being elected as Vice President of the company in 1978.

Continuing his education, Mr. McGrath received a Master's of Business Administration (MBA) degree from Iona College in New York in 1980. The following year, he was promoted to Senior Vice President and, in 1982, Executive Vice President. In 1987, Mr. McGrath became one of the trustees of the company.

In 1989, he attended the six-week advanced management program at Harvard University and was subsequently named President and Chief Operating Officer of the company. In September 1990, he accepted the position of Chairman of the Board and Chief Executive Officer of ConEd.

Raymond McGuire

Raymond (Ray) McGuire joined Kauff McGuire & Margolis LLP (KM&M) in 1981 and assumed primary responsibility for its litigation practice in federal and state courts, as well as before the National Labor Relations Board (NLRB) and other federal agencies. Since 1981, he has tried numerous cases in state and federal courts in the New York metropolitan area involving employment contracts, discrimination, and labor relations issues.

Mr. McGuire is also responsible for the firm's representation of its construction industry clients in all aspects of their labor relations, serving as Managing Director and General Counsel to the Contractors' Association of Greater New York, Inc. (CAGNY), a multi-employer association of high-rise construction managers, as well as the Association of Contracting Plumbers of Greater New York.

Figure A4.23
Raymond McGuire.

Mr. McGuire has particular expertise also in the labor issues arising in the voice and data communications industries and has successfully guided clients in these industries with respect to the labor relations implications of mergers, acquisitions, and reorganizations. He has also negotiated throughout the United States with scores of Communications Workers of America (CWA) and International Brotherhood of Electrical Workers (IBEW) local unions, including locals in California, Chicago, Houston, Dallas, and New York. He has also serviced clients in New York's maritime sector, negotiating with the International Longshoremen's Association (ILA), Seafarers International Union (SIU), and the International Organization of Masters, Mates & Pilots (MM&P), and also in New York's hospitality industry that deals with the Service Employees International Union (SEIU) Locals 6 and 100.

Raymond McGuire received his undergraduate degree from Canisius College and his law degree from Harvard Law School. Following practice with a labor law firm in New York, he served in the Peace Corps in Nigeria for two years and then obtained a Master of Law degree from Columbia Law School. He served as a professor of law at the University of Maine School of Law from 1973 to 1981, and also taught and lectured in labor and employment law at Florida State University and University College Dublin.

Robin Panovka

Robin Panovka co-chairs Wachtell, Lipton, Rosen & Katz's Real Estate and Real Estate Investment Trusts (REIT) mergers and acquisitions (M&A) practices, which are at the forefront of M&A and other major transactions in the REIT, private equity, real estate, hospitality, and gaming industries. He also advises on activism defense, strategy, cross-border transactions, and large-scale development projects, including the redevelopment of the World Trade Center and Little Island in Manhattan.

Mr. Panovka writes and lectures widely, including co-authoring *REITs: Mergers and Acquisitions*, a leading treatise published by Law Journal Press (2023). He is Co-Chair of the New York University REIT Center and has served as an Adjunct Professor at Columbia Business and Law Schools and in NYU's Schack Institute Real Estate Program. He is a Founding Director of the International Institute for the Study of Cross-Border M&A (XBMA), a joint venture among Peking University, the University of Cambridge, and NYU.

Figure A4.24
Robin Panovka.

Robin Panovka was heavily involved in the redevelopment of the World Trade Center for more than a decade following its destruction on September 11, 2001, and played a hand in negotiating the Master Plan and "footprint swap," which paved the way to rebuilding, as chronicled in the *Cornell Real Estate Review*, *American Lawyer* magazine, and other publications. He is a graduate of Cornell University and the Duke University Law School.

Figure A4.25
Steve Plate.

Steve Plate

Steve Plate is Chief of Major Capital Projects at the Port Authority, a position he has held since 2015. Prior to that, Steve served as the Deputy Chief of Capital Planning and Director of World Trade Center Construction.

As Chief of Major Capital Projects, Mr. Plate leads the most critical Port Authority megaprojects, which total more than $65 billion, including the Port Authority Bus Terminal replacement, the PATH-to-Newark Liberty International Airport extension, the Bayonne Bridge "Raise the Roadway" Navigational Clearance Project, and the emerging Gateway Tunnel Project.

He attended Manhattan College, where he earned his Bachelor's degree in civil engineering. Mr. Plate is also active in the community, having served as Mayor of Glen Ridge, NJ, where he has lived with his family since 1981.

Figure A4.26
William Plunkett.

William Plunkett

William Plunkett is a Senior Counsel of Plunkett PLLC. Over the past four decades, he has built a successful practice with Plunkett & Jaffe PC, which later merged with McKenna Long & Aldridge LLP and then with Dentons US LLP.

Mr. Plunkett has litigated cases in every state and federal court in the State of New York. He has worked with virtually every NYS agency and authority and has provided advice and counsel to Fortune 500 CEOs and leaders. He has provided counsel on major commercial and residential real estate and economic development projects in New York, including several billion-dollar projects such as the post-9/11 World Trade Center redevelopment.

Figure A4.27
Joshua Ramus.

Joshua Ramus

REX's Founding Principal, Joshua Ramus, leads the firm's think tank of diverse professionals. In 2016, he was the first American recipient of the Marcus Prize, the $100,000 biennial international architecture award conferred by the Marcus Corporation Foundation and the University of Wisconsin. He has also been credited as one of: the "5 greatest architects under 50" by *HuffPost*; the world's most influential young architects by *Wallpaper**;* the 20 most influential players in design by *Fast Company*; "The 20 Essential Young Architects" by *ICON* magazine; and the "Best and Brightest" by *Esquire*. In 2020, Mr. Ramus was honored by the experimental performance company STREB with their Action Maverick Award.

Joshua Ramus has been an Eero Saarinen Visiting Professor at Yale University, a Cullinan Visiting Professor at Rice University, and a Visiting Professor at Columbia University, Cooper Union, Harvard University, Massachusetts Institute of Technology, and Syracuse University. He holds a Master of Architecture degree from Harvard University, where he earned the inaugural Araldo Cossutta Fellowship and the SOM Fellowship, and a Bachelor of Arts in philosophy, *magna cum laude* with distinction in the major, from Yale University. He is NCARB Certified and a registered architect in states throughout the United States and Australia as well as in the Netherlands.

Mr. Ramus was a Founding Partner of OMA New York in 2001 and remained its Principal until he rebranded the firm as REX in 2006. In that time, he was Partner-in-Charge of all the firm's projects, including the Guggenheim-Hermitage Museum in Las Vegas and the Seattle Central Library, hailed by *The New York Times* architecture critic, Herbert Muschamp as "the most exciting new building it has been my honor to review in more than 30 years of writing about architecture." His most recent accomplishment is the design and September 2023 debut of the Perelman Performing Arts Center at the WTC.

Scott Rechler

Figure A4.28
Scott Rechler.

Scott Rechler is CEO and Chairman of RXR Realty (NY). He is the former Vice Chairman of the Port Authority of New York and New Jersey and, until recently, Chairman of the Regional Plan Association (RPA), and a member of the Metropolitan Transportation Authority (MTA) board from 2017 to 2019.

Scott Rechler's early career was with the family business Reckson Associates Realty Corp. (NYSE: RA). Mr. Rechler became Chief Operating Officer and later Co-CEO of the company, and he led Reckson's expansion beyond Long Island into the New York and tri-state region. In 2003, he was named Chief Executive. The company was acquired in early 2007 by SL Green Realty Corp. for $6 billion. Afterward, Mr. Rechler formed RXR Realty and became CEO and Chairman of the organization.

Mr. Rechler also serves as trustee and Vice Chairman of the National September 11 Memorial & Museum at the World Trade Center, is a member of the Real Estate Board of New York (REBNY), and of the Real Estate Roundtable (RER), a national organization. He is Chairman of the Feinstein Institutes for Medical Research, a member of the board of trustees at Northwell Health, and the Hospital for Special Surgery, and a member of the NYU Real Estate Institute Advisory Committee.

Scott Rechler is a graduate of Clark University and the New York University Schack Institute of Real Estate.

Figure A4.29
Raymond E. Sandiford.

Raymond E. Sandiford

Raymond ("Ray") Sandiford is the national practice lead for geotechnical engineering for HNTB. He has 45 years of experience in civil design and heavy construction. His experience has primarily been in the transportation industry, and he has led numerous port facility, aviation, rail, and roadway projects.

Prior to HNTB, Ray Sandiford was the Chief Geotechnical Engineer for the Port Authority, where for 30 years, he was involved in all the major development work at all the agency's facilities. He brought much of his previous Port Authority and private experience to the WTC's 9/11 recovery and its redevelopment.

He is a licensed professional engineer in the states of New York, New Jersey, and Virginia, and he received his Bachelor's degree from the State University of New York (SUNY) at Stony Brook and his Master's degree from Columbia University. He is a member of the Society of Manufacturing Engineers (SME), an emeritus member of both the Moles and the American Society of Civil Engineers (ASCE) and is a Visiting Lecturer on soil mechanics at Princeton University.

Figure A4.30.
Frank Sciame.

Frank Sciame

Frank J. Sciame Jr. is Chairman and CEO of Sciame Construction Company. He established the company in 1975, followed by the founding of Sciame Development in 1996. Mr. Sciame's work is particularly well known in NYC, where he often is referred to as "the builder of choice." Projects include some of the most notable landmarks in NYC, including the Morgan Library, the New Victory Theatre, the New York Public Library's Main Branch, and the Central Synagogue restorations.

Sciame Construction has just completed the Perelman Performing Arts Center (PAC) at the WTC; has completed the Morgan Library & Museum expansion and exterior restoration; the Studio Museum in Harlem; the Frick Collection expansion; the New Museum expansion; the Shed at Hudson Yards; the Museum of Jewish Heritage in NYC; and the Polonsky Shakespeare Center in Brooklyn.

Mr. Sciame has given back to multiple charitable, civic, and professional organizations. For his dedication to the preservation of NYC landmarks, he received the Pillar of New York Award from the Preservation League of New York State (2003) and the Chairman's Award from the New York Landmarks Conservancy (1997). For his public service, he was presented with the Cardinal's Committee for Charity Award by Edward Cardinal Egan. In November 2023, he was awarded the Jacqueline Kennedy Onassis Medal from the Municipal Art Society. He holds a Bachelor's degree in architecture from the City College of New York and an honorary Doctor of Fine Arts from CCNY.

Christopher O. Ward

Christopher ("Chris") O. Ward is currently Executive Vice President for Business Development at Bravo Inc., an architectural and engineering company which is a Minority and/or Women Owned Business Enterprise (MWBE) . Prior to that, he was Executive Vice President and Chief Executive Officer of metro New York for AECOM, a $19 billion, fully integrated infrastructure and support services firm. Prior to joining AECOM, Mr. Ward served as Executive Vice President for major projects at Dragados USA, where he was responsible for significant business development throughout the United States.

Chris Ward's 30+-year career in infrastructure and government comprises experience in both the public and the private sectors, including a successful tenure as Executive Director at the Port Authority of New York and New Jersey, where he led the World Trade Center rebuilding effort and executed the turnaround at the site by implementing creative engineering solutions—delivering on a commitment to open the 9/11 Memorial by the 10th anniversary of the tragic events at the site. Earlier in his career, Mr. Ward served in the first term of the Bloomberg administration in NYC, leading the Department of Environmental Protection in a robust expansion of its capital plan.

Mr. Ward earned his Master of Theological Studies degree at Harvard University and his BA in history and religious studies at Macalester College.

Figure A4.31
Christopher O. Ward.

Carl Weisbrod

Carl Weisbrod is a national authority on city planning, affordable housing policy, urban development, and public–private real estate development. In his four decades of experience, he has brokered and led complex public–private partnerships, and created innovative governance strategies to spur economic development. His record of visionary, innovative, and successful urban development initiatives has made him an internationally sought-after advisor.

Carl Weisbrod has served the City of New York in many capacities—most recently, as Chairman of the New York City Planning Commission and Director of the New York City Department of City Planning. Prior to these appointments, he was a partner at HR&A, where he managed the successful rezoning of the Hudson Square area in Manhattan. His former positions include President of New York State's 42nd Street Development Project, founding President of the New York City Economic Development Corporation, founding President of the Alliance for Downtown New York, and President of the real estate division for Trinity Church, and Executive Vice President of Trinity Church-St. Paul's Chapel.

Carl Weisbrod was the inaugural recipient of the 2017 John E. Zuccotti Public Service Award from the Real Estate Board of New York, which recognizes citizens who have displayed exceptional accomplishments and service in the public's interest. In 2011, he was recognized by the Center for an Urban Future as "one of New York's most innovative economic development leaders over the past three decades," and in 2002, he was honored by *Crain's New York Business* as one of the "100 Most Influential Leaders in Business." He has received both the Chairman's Award from the New York Landmarks Conservancy (2000) and the Public Interest Service Award from New York University (1993).

Figure A4.32
Carl Weisbrod.

Carl Weisbrod holds a Bachelor of Science degree from Cornell University and a *Juris Doctor* (JD) from New York University School of Law. He is a board member of the Lower Manhattan Development Corporation, a Senior Fellow of the Marron Institute of Urban Management, and a board member of the Trust for Governors Island, New York.

Figure A4.33
Joe Woolhead.

Joe Woolhead

Born and raised in Dublin, Ireland, Joe Woolhead has been living in NYC for over 30 years. Initially, he worked in construction as a laborer, painter, plasterer, and stonemason. After a serious accident at One Chase Manhattan Plaza in 1996, he returned to college and rediscovered his passion for creative arts, particularly poetry and photography. He graduated from Hunter College with a degree in film in 2002. When the attacks happened on September 11, 2001, he grabbed his camera and shot what he witnessed over the course of three days. His photographs from that time were subsequently published all over the world.

In 2004, Joe Woolhead began working with Silverstein Properties, focusing first on the rebuilding of 7 WTC. In 2006, he started work at the 9/11 Memorial, documenting the construction of the Museum and the Memorial Pools. He continued his work at 1 WTC, 4 WTC, 3 WTC, the Oculus, and the Perelman Performing Arts Center (PAC), ultimately shooting over 3 million photographs of the iconic WTC project. In his role as official World Trade Center photographer, Mr. Woolhead collaborated with author Scott Raab on *Once More to the Sky* (Simon & Schuster, 2021), for which he also wrote the prologue. Joe was also featured prominently in a book by Matt Kapp, *A Century Downtown* (2020), chronicling his three days spent at Ground Zero in 2001. Joe Woolhead lives in Queens with his wife, Ozy, and their son, Chet.

Figure A4.34
Tom Wright.

Tom Wright

Tom Wright is President and Chief Executive Officer of the Regional Plan Association (RPA), the nation's oldest independent metropolitan research, planning, and advocacy organization. A private, nonprofit corporation, RPA improves the prosperity, infrastructure, sustainability, health, and quality of life of the New York-New Jersey-Connecticut metropolitan region by researching and preparing long-range plans and advocating for their implementation.

As President, Tom Wright led the production of RPA's landmark Fourth Regional Plan, released in 2017, which proposed 61 recommendations to reform public institutions, modernize transportation systems, tackle the challenge of climate change, and provide affordable and livable communities for all the region's residents. RPA is now working to implement the major ideas in the plan, such as charging all drivers to enter the Manhattan CBD, cutting carbon emissions, and scaling up renewable energy sources; creating healthy, affordable housing in every community; modernizing the NYC subways, and building a new commuter rail tunnel under the Hudson River connected to a renovated and expanded Penn Station.

In his 20+ years with the organization, Mr. Wright has participated in many key RPA initiatives, including the historic Civic Alliance to Rebuild Downtown New York following the September 11, 2001, attacks; the campaign to create a mixed-use district at

Manhattan's Hudson Yards; the protection of the New Jersey Highlands; a vision plan for the city of Newark; and The Constant Future, a public exhibition in Grand Central Terminal's Vanderbilt Hall that celebrated RPA's centennial. He was also in charge of producing *A Region at Risk*, RPA's third plan published in 1996.

Previously, Tom Wright was Deputy Executive Director of the 2001 New Jersey Office of State Planning, where he coordinated the production of the New Jersey State Development and Redevelopment Plan. In January 2020, Governor Phil Murphy appointed Mr. Wright as the Chairman of the New Jersey State Planning Commission.

Mr. Wright is a frequent public speaker and commentator on regional governance, economic growth and development, transit investments, and other policy issues. A Visiting Lecturer at the Columbia University Graduate School of Architecture, Planning and Preservation, he has a Master's of Urban Planning from Columbia University, a Bachelor's degree in history, and a certificate in American studies from Princeton University.

Index

O

P

R

S

T

W

Z